U0350558

Aedas

IN CHINA 在中国

支文军 ZHI Wenjun 徐 洁 XU Jie 编著

序 1
Preface 1

文 郑时龄 by ZHENG Shiling

在这个设计公司和事务所林立、建筑大师云集的时代，1985 年才在中国香港成立的 Aedas 纯属年轻一代。Aedas 从中国香港和东南亚起步（1985—2000），2000 年起陆续转向在中国大陆、中东和欧洲的发展，2002 年首先在北京开办业务，2005 年在中国澳门、上海和成都开办事务所，至今已在伦敦、中国香港、北京、成都、中国澳门、上海、深圳、新加坡、新德里、西雅图、阿布扎比、迪拜等地都建立了事务所，所从事的业务也从建筑设计拓展到城市规划、城市设计、室内设计、景观设计等综合的领域。Aedas 在中国的设计遍及北京、上海、广州、深圳、珠海、中国香港、中国澳门、中国台北、成都、重庆、大连、长沙、武汉、青岛、无锡、苏州、义乌等城市。Aedas 的设计涵盖了办公楼、商业中心、文化中心、酒店、交通建筑、学校、住宅以及建筑改造等多种类型，尤其擅长复合功能的综合设计，近年来又向文化和艺术建筑拓展，作品获得奖项之多令人咂舌。这家新生代的跨国建筑设计公司在短短 30 多年间的发展和斐然的成绩令人刮目相看。

一家大型跨国建筑设计公司的成功与否取决于公司的作品、商业运作和公司文化。我认识 Aedas，缘于一位在 Aedas 上海事务所工作多年的同济大学毕业生的介绍，我也曾经遇见他们的设计方案，尤其是参与讨论过上海星荟中心的项目，参观过由 Aedas 设计的苏州西交利物浦大学中心楼，因而对 Aedas 的设计留有深刻的印象。当今，建筑师所从事的工作是综合了社会性、艺术性、技术性、逻辑性和创造性的理性活动，是从现实世界迈向未来世界的一种创造，也是一种想象的飞跃。历史上将建筑师与医生、律师并列为三大古老的职业。在人们的概念中，医生维护人体的秩序，律师维护社会的秩序，而建筑师则建构物质世界的秩序，甚至是未来世界的秩序。建筑作品与建筑师的品格及其鉴赏力密切相关，好的建筑是因为有好的品味。英国艺术评论家罗斯金（John Ruskin，1819—1900）曾经说：“愚蠢者用愚钝的方式建造，智者敏锐地建造，善良的人建造美丽的建筑，而邪恶的人建造卑劣的建筑。”

在当今的建筑建设领域，成功不仅要依靠商业运作，更需要成功的设计，“以设计说话”“以成败论英雄”也许是这个领域普遍的竞争法则。这就需要人才，需要公司的领导力，需要社会的培育。Aedas 之所以能获得众多城市的认可，说明社会能接受 Aedas 建筑师的创造。Aedas 的设计具有多样性，在他们的设计中我们可以看到一种创造力和追求时空变化的倾向，也许与时下流行的建筑并不相似，但是正如斯洛文尼亚建筑师米卡·奇莫利尼（Mika Cimolini，1971— ）所说：“创造性的本质就是挑战传统空间，并创造新的空间。创造就是为建筑的使用者激发新的文化体验。”事实上，创造性应当以历史的尺度来衡量，而不是单纯以当下的技术进展作为参照，因此，创造性是面向未来的。中国哲学历来视时空为统一体，宇宙、世界是时空的统一。我们通常所说的宇宙，就是“四方上下曰宇，往古来今曰宙。”而我们所说的世界，世即宙，即时间；界即宇，即空间。从 Aedas 的设计中，我们可以看到一种根植在中国传统文化中的融合时间和空间的创作意图，相信时空的融合会使这家年轻的跨国建筑设计公司更加快速地成长。

如果今天仍然可以将建筑比喻为凝固的音乐的话，那么 Aedas 演奏的是现代复调音乐。

郑时龄 | ZHENG Shiling

中国科学院院士，同济大学教授，意大利罗马大学名誉博士，法国建筑科学院院士，美国建筑师学会荣誉资深会员，同济大学建筑与城市空间研究所所长，上海市规划委员会城市发展战略委员会主任委员。

Member, the Chinese Academy of Sciences. Professor, Tongji University; Laurea honoris causa, Università di Roma, La Sapienza; Membre de l'Académie d'Architecture de France; Honorary Fellow, the American Institute of Architects; Director, Institute of Architecture & Urban Space, Tongji University; Director, the Committee for Urban Development Strategy, Shanghai Planning Commission.

Compared with many of the leading design firms operating today, Aedas is rather young. Founded in Hong Kong in 1985, it expanded into Southeast Asia and then, after 2000, into Chinese mainland, the Middle East and Europe. It opened an office in Beijing in 2002, followed by Macao, Shanghai and Chengdu in 2005. Aedas currently has offices in London, Hong Kong, Beijing, Chengdu, Macao, Shanghai, Shenzhen, Singapore, New Delhi, Seattle, Abu Dhabi and Dubai. Its business has expanded from architectural design to master planning, urban design, interior design, landscape design and more. In China, Aedas' projects can be found all over, covering Beijing, Shanghai, Guangzhou, Shenzhen, Zhuhai, Hong Kong, Macao, Taipei, Chengdu, Chongqing, Dalian, Changsha, Wuhan, Qingdao, Wuxi, Suzhou, Yiwu and more. Aedas' designs run the gamut from office buildings and shopping malls to theatres, cultural centers, hotels, airport terminals, transportation infrastructure, schools, residential projects and renovations. With strong expertise in multifunctional complex design, Aedas has won many awards. It may have been around for just 30 years, but this young, multinational architectural design company has already delivered impressive results.

As a large multinational architecture firm, success depends on the quality of its works, the smoothness of its business operations and the healthiness of the company culture. I have been familiar with Aedas for many years. At first, it was introduced to me by a graduate from Tongji University who worked in Aedas' Shanghai office. As I became familiar with Aedas' works, I was deeply impressed, especially after visiting the Shanghai Landmark Center and Xi'an Jiaotong-Liverpool University Central Building. Today's architects are engaged in work that combines sociality, artistry, technology, logic and creativity. It's a practice that derives the future world from what exists in the present. It requires a great leap of imagination. Architects, doctors and lawyers are three of the oldest professions in history. Doctors keep the human body in good working order, lawyers protect the social order, while architects shape the order of the physical world — and even the order of the future. Architectural works are closely related to architects' character and taste. British art critic John Ruskin (1819-1900) once remarked, "A foolish person builds foolishly, and a wise one sensibly; a virtuous one, beautifully; and a vicious one, basely."

In architecture, success is about business, but also the quality of design. Design talks. Success talks. These are the general rules of competition in this field. This requires talent, company leadership and social nurturing. That Aedas has been able to gain recognition from so many cities suggests its architects have been embraced by society. In Aedas' diverse designs, we can see the ambition to pursue creativity and change in time and space. That may not reflect the most popular trends in architecture, but as Slovenian architect Mika Cimolini (1971-) has said, "The essence of innovation lies in creating spaces that challenge common patterns of usage and find new ones. In fact, creativity should be measured on a historical scale rather than purely by the progress of current technology. In this sense, creativity is oriented toward the future. Chinese philosophy regards time and space as one unified reality; both the cosmos and the world are the combination of time and space. The Chinese word for 'world' is formed by two characters implying time and space, respectively. We can see the creative intention to integrate time and space in Aedas' projects. I believe that intention will enable this young international firm to grow quickly.

Architecture has been compared to "solidified music." If that's true, then what Aedas plays is modern polyphony.

序 2
Preface 2

文 纪达夫 by Keith GRIFFITHS

Aedas 自 1985 年成立以来，经过不断成长，逐渐获得了全球设计界的认可。为了理解其原因，我们须考虑过去 30 多年的全球经济以及 Aedas 的实践理念和合作架构。

Aedas 于 20 世纪 80 年代中期成立于中国香港，与八九十年代迅猛发展的东南亚“老虎”经济一同发展。当时，东南亚城市在商业、住宅和基础设施建设发展方面，尤其是预制建筑技术领域亟需要国际级设计专业人才。

作为亚洲为数不多的、具有预制建筑设计经验的国际级设计团队， Aedas 承接到大量当地商业和基础设施项目。 特别是 Aedas 在与香港铁路有限公司的合作中取得了轻轨建设的经验，有助于 Aedas 于 20 世纪 90 年代在新加坡和吉隆坡设立分公司。到 2000 年，Aedas 在东南亚的四个公司已拥有 250 名员工。

高层建筑设计方面的专业性保证了 Aedas 在全球各地的交流项目顺利实施。2005 年之前，Aedas 在伦敦、迪拜、洛杉矶和西雅图分别设立了分公司。2002 年 Andrew BROMBERG 的到来提高了 Aedas 设计的国际声誉，Andrew BROMBERG at Aedas 的建立也是对他在设计实践的领导力的肯定。

新千年为国际建筑师们带来了更多机会参与到中国的经济增长中。从 1990—2017 年的 27 年间， 中国城市化进程涉及的人口超过 4 亿。 在项目所在地建立办公室的做法进一步巩固了 Aedas 的成功，2005 年，Aedas 在北京、上海、成都和中国澳门设立了办公室，接着又在广阔的中国市场采用了同样的模式。12 年后，Aedas 的 12 个办公室已拥有 1 400 名员工，在中国为当地项目工作的员工超过千人。

Aedas 的成功很大程度上归功于亚洲经济的迅速发展，这也造就了适应各城市的地域文化、地理特点的优秀设计。 毫无疑问，设计实践的稳定和成功也可归功于 Aedas 跨国际的合作架构体制及其全球视野。

为了提高设计标准，开拓业务发展，Aedas 通过股权来激励 30 位设计总监。 其中 9 位国际设计总监设定了设计实践的总体目标、标准和国际化的操作平台。 所有 Aedas 股东都是全职的建筑设计师，而实践管理和项目由相应的专业人员和执行建筑师来处理。 每个项目都需要一位设计总监和一位联合设计总监操刀，两者尽量来自不同地区的办公室。 共享的设计领导力确保了专业设计知识和想法的自由流通，设计过程完全透明。

中国正在打造世界上规模最大、密度最高的城市，这些城市都将成为智能技术、土地和能源效率及交通协作的典范。 新的建筑形式是对新新人类的回应，他们有着新的居住办公和信息获取的方式。 中国的城市不受旧的基础设施和建筑的阻碍，必将引领世界，成为未来的城市形态。 Aedas 在亚洲有着 30 多年的从业经验， 是为数不多的、有能力向世界推广高密度城市新模式专业知识的国际建筑设计公司。

纪达夫 | Keith GRIFFITHS

纪达夫是 Aedas 的主席及全球设计董事。Aedas 是全球最大的建筑设计事务所之一，在遍布亚洲、美洲和欧洲的 12 个办公室拥有共 1 400 名创意人才。纪达夫自 1983 年起居于香港。
作为享誉世界的建筑师及城市规划师，纪达夫拥有 35 年丰富的设计经验，作品涵盖城市规划，以及高层、高密度的综合发展项目、机场、轨道和市政设施等各类项目设计，地域遍布亚洲各地。

Keith GRIFFITHS is the Chairman and Global Design Principal of Aedas, one of the world's largest architectural practices with 1,400 staff in 12 offices throughout Asia, USA and Europe. He has lived in Hong Kong since 1983.
Keith is an internationally respected architect and urban planner with 35 years experience in city scale urban planning and in the design of high-rise, high-density developments, airports, rail and civic facilities throughout Asia.

Aedas has achieved global design recognition and spectacular growth since its formation in 1985. To understand the reasons for this we must consider both the global economy of the last 30 years as well as the practice's philosophy and collegiate structure.

Aedas was born in Hong Kong in the mid-1980s and grew alongside the rapidly developing Southeast Asian 'Tiger' economies of the 80s and 90s. The cities of Southeast Asia required international design expertise in commercial, residential and infrastructure developments and in particular in prefabricated building technology.

Being one of the few international practices in Asia with experience in the design of prefabricated buildings enabled Aedas to attract commissions upon both commercial and infrastructure work throughout the region. In particular Aedas acquired expertise in rail from work with the MTRC in Hong Kong and this helped to establish offices in Singapore and Kuala Lumpur in the 1990s. By 2000 the practice had grown to 250 staff in its four Southeast Asian offices.

The practice's expertise in high rise design positioned it well to secure global commissions and to establish further offices in London, Dubai, Los Angeles and Seattle by 2005. Aedas' international design reputation was firmly secured by the arrival of Andrew BROMBERG in 2002 and the establishment of Andrew BROMBERG at Aedas in recognition of his design leadership of the practice.

The new millennium brought greater access for international architects to participate in China's economic growth which, in the 27 years from 1990 to 2017 urbanised in excess of 400 million people. Aedas' success had been underpinned by establishing local offices and gaining a thorough understanding of local requirements allied to international expertise. The practice decided to follow the same model for the vast China market with offices in Beijing, Shanghai, Chengdu and Macao operating by 2005. 12 years later Aedas had grown to 1,400 staff in 12 offices with over a thousand of the staff being employed in China upon China commissions.

Whilst the success of Aedas is largely due to the rapid expansion of the Asian economy, this was underpinned with excellent design appropriate to the culture and geography of each city and locality. Undoubtedly the stability and success of the practice can also be credited to its collegiate structure and ownership which straddles both international and global perspectives.

30 Design Directors are incentivised through their local ownership to drive the high design standards and business. Nine International Design Director owners set the overall goals, design standards and international operational platform of the practice. All of Aedas' owners are full time architectural designers whilst the administration of the practice and its projects are handled by appropriate professionals and administrative architects. Each project requires a Design Director together with a Co-design Director preferably from another office. This sharing of design leadership ensures a free flow of design knowledge and ideas throughout the organisation and complete transparency of the design process.

China is now developing the world's largest and highest density cities which are models of smart technology, land and energy efficiency and transport coordination. New building forms respond to a new live-work, information savvy population. Unhindered by old infrastructure and buildings, China's cities will lead the world as the urban form of the future. With 30 years experience in Asia, Aedas is one of the world's very few international architectural practices with the expertise to bring to the world the new model of high-density, high-rise city.

目 录

Contents

Aedas 设计理念综述

Overview of Aedas Design Philosophy

文 支文军 by ZHI Wenjun

Aedas 发展历程

Aedas，源于拉丁语 aedificare，即“to build”：“建筑”之意。

自 1985 年 9 月 Keith GRIFFITHS（纪达夫）最初在香港设立设计工作室至今，Aedas 始终关注于打造开放包容的设计事务所，并为每位加入的设计同僚提供能够鼓舞人心、激发积极性和认同感的工作环境。Aedas 从一个仅有两人的设计工作室发展为以设计品质闻名全球的设计公司，始终保持着对市场热点的敏感度，以市场需求引导公司发展路径，至今已经历了 33 年。根据各个时期的不同核心战略，大致可将 Aedas 的发展历程划分为四个阶段：

20 世纪 80 年代——主要位于中国香港，以高层为主的高密度建筑

20 世纪 80 年代，中国内地的城市化进程尚未开始，Aedas 的实践主要位于中国香港。此时欧洲面临经济危机，而香港正处于经济上升期，全球化的浪潮给香港转型带来了难得的机遇，城市飞速发展，一跃成为“亚洲四小龙”之一，更成为了亚洲的区域核心。

这一时期的香港建筑以砖、混凝土结构建筑为主。伴随着办公楼、购物中心的发展，香港开始需要更为复杂的建筑形式，如标准化构件、预制装配式建筑等。Keith GRIFFITHS 以国际化的视野在香港创立了工作室，结合来源于欧洲的专业设计水平，以多样化的创造性设计和工艺，灵活应对这一变化需求。

20 世纪 90 年代——扩展至东南亚区域，转向定制化建筑

20 世纪 90 年代，Aedas 的发展着力点在中国香港和东南亚。新加坡、马来西亚、印度尼西亚等东南亚国家的经济繁荣，使得 Aedas 由此获得了更多设计基础设施、住宅和商业综合体的技能与经验，同时开拓出不同国家和城市的业务，能够同时为不同业主提供全面的专业服务。

Aedas 在轨道和机场设计方面的专业能力使得公司发展成为横跨东南亚和中东地区的大型企业。Aedas 与客户共同工作，根据他们的需求量身设计定制的住宅、火车站、机场等建筑。这时，Aedas 已经从设计装配式建筑转向设计定制化建筑。

2000 年至 2005 年——经历了全球化及 Aedas 的品牌重塑

2000 年至 2005 年，Aedas 积极投身于国际化竞争，更加关注于世界全球化发展。中国自 20 世纪 90 年代经济开始崛起，开始复杂的商业综合体开发。在世界范围的金融危机背景下，Aedas 以极具前瞻性的视野，主动参与到中国内地的城市建设中。

Aedas 于 2002 年在北京设立了第一家中国大陆办公室，并开始了中国大陆的第一个重要商业综合体项目。公司定位的更新与品牌的重塑，使 Aedas 得到迅速成长，并逐步成为真正的国际化公司。

2005 年至 2017 年——聚焦中国建筑发展，关注当地市场与文化

2005 年后，基于从中国香港、新加坡、吉隆坡、曼谷等快速发展的城市中所获得的经验，Aedas 更加聚焦于中国的城市发展进程，并意识到仅仅专注于国际化、全球化是远远不够的。

中国的城市间存在着奇妙的差异，这种差异促使 Aedas 主动加深对本地市场和中国文化的理解，Aedas 在“全球化”(global) 的基础上，将“当地化”(local) 加入到设计理念之中，并尝试塑造出能将“全球化”与“当地化”有机结合的、因地制宜的设计。

Aedas 的哲学

在不同的发展阶段，Aedas 会根据具体的实际情况去设定不同的目标，正如创始人 Keith GRIFFITHS 所说：“去做别人所未

支文军 | ZHI Wenjun

同济大学建筑与城市规划学院教授，《时代建筑》杂志主编，中国建筑学会建筑传媒学术委员会副主任委员。

Professor of the College of Architecture and Urban Planning, Tongji University; Editor-in-chief of *TIME + ARCHITECTURE* Journal; Vice-director of Media and Communication Committee, Architecture Society of China.

尝试过的事，因为我们可以将它做得更好。” Aedas 将多元化的理念、创新性的思想、国际性的视野融入设计，同时深度挖掘当地社会与文化，协同融汇世界优秀思想，逐步实现了各个阶段的愿景。

多元性 创新性 国际性

Aedas 坚持以多元化的思想进行因地制宜的设计，以富含创造力和创造性思维的设计满足全世界不同城市的需求，多元性、创新性、国际性共同筑就 Aedas 的独特价值观。

1）多元性

“多元化的世界是一个更加缤纷多彩的世界”，Aedas 坚信“伟大的设计即多元化的设计”。在设计中，Aedas 不断挑战现状，试图寻找走在时代前沿的思维理念和更多的可能性。

Aedas 的实践涉及建筑设计、文化艺术设计、平面设计、室内设计、景观设计、城市设计及总体规划等领域，设计类型涵盖艺术和休闲设施、基础设施、会议和展览设施、综合体、企业办公、研究和制药设施、教育设施、住宅、酒店、商业零售等。每个项目无论规模大小，都融汇了 Aedas 全球各地的智慧。

2）创新性

“Aedas 将创造力和创造性思维置于首要位置。” Aedas 将全世界的专家们集合在一起，以多样性的价值观为使用者定制有机的解决方案。对于每一个新项目，Aedas 都在不断地寻找突破和新的可能性，独特的全球服务平台为遍布世界各地的专业建筑师提供最新的创意理念。

富有灵感、有独立性思维、对设计有激情并渴望超越的设计师们是 Aedas 的核心力量，这些设计人才为服务的地区提供能产生积极影响的解决方案，Aedas 也为他们提供充分的自由空间以及所需的支持、信息和决策力。

3）国际性

Aedas 是全球唯一将国际研究、本地知识和环球业务结合起来，既扎根本土又立足世界的建筑设计公司。自 2002 年品牌重塑后，Aedas 始终保持国际水平及全球视野，不断自我提升，以适应自身发展和世界变化。

Aedas 在中国香港、北京、上海、伦敦、迪拜、西雅图等地都设置了办公室，1 400 余名创意人才汇聚于遍布全球的 12 个设计工作室，以国际化的视野与全球领先的设计水准，自觉参与到全世界城市发展的进程中。

社会与文化

“建筑连接人群，建筑连通文化”。Aedas 坚信“只有对当地社会与文化具有深刻了解的建筑师才能做出杰出的设计”。Aedas 将国际视野与地方文化相结合，在设计中联系当地语境，充分尊重当地文化。

同时，Aedas 始终贯穿“生命亲和”这一全球性发展与地域性实践的思想，也将生态可持续的理念充分融入设计之中。Aedas 将人类和自然的紧密联系与日益密集的世界城市化背景相融合，赋予了“钢筋水泥丛林”城市全新的、更加积极的意义。

团队协同

作为首屈一指的国际化建筑设计公司，Aedas 将“全球化”与“当地化”融合，使各个董事的专长与知识紧密相连，确立了一套有效的“学院式”合伙人所有制体系，再经由协同交流平台将整个 Aedas 黏合到一起。

1）全球化与当地化

Aedas 始终坚持以全球化的标准提供当地化的服务，尊重不同文化和群体的社会与环境差异。为了有效融合来自世界各地的优秀思想，Aedas 设立了协同工作的全球化平台，建立起多元化的工作关系。这个平台能够帮助设计师将设计思想发散出去，再收敛回来，最终形成最佳的设计方案。

同时，Aedas 全球设计工作平台也能够让不同办公室和国家的设计师分享创作过程，向所有 Aedas 设计师提供内部共享的软件、研发中心和绘图资源。

2）团队合作体系

目前在全球层面 Aedas 由 12 位全球董事共同管理，创建并遵守严格标准并监督其实施。每一个办公室都有当地董事的参与，这保证了 Aedas 的客户能够直接同项目董事进行合作。来自全球不同办公室的设计师，以国际平台结合当地文化，形成了 Aedas 独特的协同设计模式，项目融合了世界各地设计专家的集体智慧，使得 Aedas 的设计既国际化又“接地气”。

结合协同设计平台，每个项目通常会由 2 位来自不同办公室的设计董事负责，在项目进行过程中，不仅可以分享各自的理解与认知，也保证了 Aedas 最优秀的国际化创意和知识能够得到广泛应用。

3）协作理念

“多元”“协作”“创新”“革新”“灵感”“热情”“适应性”“品质”“领导核心”及“团队”10 个核心概念，共同支撑构建起 Aedas 的全球协同平台。这种极具实验性的独特组织结构，使 Aedas 位于不同地点的不同办公室紧密相联并能够迅捷沟通。

通过内网平台，各办公室可分享 Aedas 全球所有新项目的最新进展。在 Aedas 强大的管理、业务拓展、设计标准、软件和财务组成的信息架构支持下，不同办公室的设计师们可以分享并交流设计信息、范例、科研文章以及最新发展成果。

Aedas 在中国

从 1985 年 9 月在中国香港设立设计工作室，到 2002 年在北京设立中国大陆第一家办公室，直至今日 Aedas 在中国建筑市场已积累了 30 余年的实践经验。Aedas 在中国的项目总数占据了其全球项目的 60%~65%，在中国大陆及中国香港的办公室拥有近 1 000 名员工。

作为一家全面的建筑设计咨询公司，Aedas 承接了相当数量的由跨国公司开发的系列项目，成功探索出“全球与当地”相结合的设计模式。除了具备国际性的专业知识及能力外，又对本土文化、城市历史文脉及绿色生态建筑累积了深刻认知，因此得以创造出因地制宜的在地性设计。

“全球与当地”

Aedas 全球化的发展依靠的是对远景强大的洞察力。2005 年以后，Aedas 在“全球化”的基础上，将“当地化”土化加入了设计理念之中，形成了“全球与当地”的独特设计理念。

1）本地知识驱动全球业务

Aedas 的使命是创造出世界顶尖的设计解决方案，因地制宜地满足全世界不同城市以及当地经济、社会与文化的需求。在全球的城市化浪潮中，建筑设计常常面对的挑战之一是混乱而缺乏特色的基地环境，但在这背后恰恰蕴藏着历史、社会和文化的丰富内涵。Aedas 发掘文化导向的设计策略，辅以可持续发展的策略，认为建筑项目既要对环境负责，也要对历史负责。

2）在地性实践

如何才能更具特色地应对中国自身的需求是 Aedas 一直在探索的课题。随着世界城市化的进程，中国正处于迫切需要发展的阶段。Aedas 极具天赋的建筑师们扎根本土，充分从中国博大精深的传统文化、独具特色的地理和社会人文风貌中汲取灵感养分。Aedas 自觉选择在建筑当地进行设计，吸纳本地的建筑师参与工作，他们对于当地历史文脉的熟知有效保证了 Aedas 在建筑市场中的本地优势。

北京财富中心第一期是 Aedas 在中国大陆的第一个商业综合发展项目
Fortune Plaza Phase 1, Beijing, is Aedas' first commercial mixed-use project in mainland China

Aedas 自 1996 年来与置地控股深度合作，将其位于香港中环核心地带的 12 座建筑连接形成人行天桥网络系统
Since 1996, Aedas has been working on the footbridge network system that links the 12 buildings by Hongkong Land in the heart of Central, Hong Kong

以 Aedas 设计的西交利物浦大学中心楼为例，该项目以苏州城附近出产的太湖石为概念，将“学者之石”的设计理念融合到“立体苏州园林”的空间体验中，同时将中国传统中实体与虚空相互流转的空间观与建筑微气候营造相结合，形成了优雅而自然的有机建筑。

城市

全球化引发了中国城市的快速发展，但同时全球化语境淡化了中国建筑和东方文化的主体意识，由此引发了建筑文化的国际化以及城市空间的趋同。Aedas 以多元化和地域化并存的思想，探索出“城市枢纽”的概念，为中国设计了众多优秀的城市综合体建筑，有效应对全球化带来的冲击。

1）城市化

在过去的 20 年间，中国有近五亿人参与了城市化进程，土地利用的压力要求城市更加集约、高效和高密度，未来的城市将如何提供一个现代、便捷、通达且可持续发展的环境，成为了 Aedas 探索的重点。通过对中国新型城镇化规划的解读，Aedas 认为城市不应再以侵吞耕地的方式盲目扩张，而应将已有的城市进行整体规划，进行全局化、精细化发展。土地的内涵不仅仅体现在财富价值，更应是人文精神价值的载体。

2）城市枢纽

Aedas 预测，未来城市将产生一系列围绕在中央商务区周边的高密度城市枢纽。这些城市枢纽具有高密度、基于现有交通节点、复合功能及高度适应性、多平台拆分、高度连接性等特点。城市将围绕中央商务区发展形成城市枢纽网。城市枢纽将成为高密度的工作、生活及休闲中心，所有设施都位于步行可达范围。

建筑物将变得具有渗透性，首层将被拆分为多个公共平台，使景观、光线和空气相互渗透贯通。新的城市枢纽网络将成为城市居民生活的全新理想归属，并为新的城市属性提供范本。

3）城市综合体

综合体的概念源自于西方，但在中国城市中得到了广泛实践。综合体不仅是传统意义上的“混合功能体”，而是一个新的城市社会生活空间模型。城市是场所的集合，城市综合体关乎生活、工作、居住、SOHO、LOFT、办公、交通、零售商业、娱乐等，各类事件与活动都在这一场所发生，是一种新型都市形态的集合。

城市综合体建筑的每个楼层都应得到充分利用，经由良好规划的公共空间，不仅与四通八达的交通枢纽相连，并在多个层面与其他建筑体相接，方便各种活动的开展，有效提高生活质量。城市综合体建筑中的不同业态组合，能形成动态多变的建筑形式。传统现代建筑受制于单一特定用途，而综合体建筑却没有这样的桎梏，因为容纳了许多不同的功能，所以更具灵活性和适应性。

Aedas 在中国设计了许多城市综合体。如成都恒大广场，建筑模拟四川自然地貌景观，设计灵感来自天然梯田池，其购物中心与城市公共空间室内外融合连接，是具有多孔型与贯通性的设计。建筑更在形式与功能之间形成了完美平衡，为充满活力的城市中心创造出一座都市绿洲。

历史文脉

Aedas 极具天赋的建筑师们扎根本土，充分从中国博大精深的传统文化和独具特色的地理和社会人文风貌中汲取灵感养分。他们将这些中国元素与别出心裁的创意相结合，创造出世界顶尖的设计方案。Aedas 一直坚持对当地社会与文化有深刻了解的设计，不只借由造型和结构表现当地特色，还须将内涵链结到建筑功能本身，所以 Aedas 始终与当地社区同心同行，试图挖掘本地更深层的历史文脉。

德国作家歌德说：“建筑是凝固的音乐”。Aedas 巧妙地运用建筑所特有的韵律与力量与使用者产生共鸣，并利用建筑设计追溯关于当地文化、历史事件和地理环境的回忆，使设计与城市紧密相连，唤起人们记忆中精彩有趣的故事并代代相传。以大连恒隆广场项目为例，Aedas 从中国的传统文化艺术中找到借鉴元素，以常见于中国农历年画及剪纸艺术中的“如意双鲤”为意向，糅合了东方韵味和现代风格，巧妙呼应了周遭环境和当地社区，把一个生动的建筑故事向使用者娓娓道来。

可持续发展

设计和创造文化与环境可持续的建筑是 Aedas 的使命。Aedas 认为，对于自然的需求是人类天性中的一部分，尽管城市居民出生并成长于密集的城市环境中，但仍应以各种方式去拥抱大自然。Aedas 内部专门设有可持续团队，努力探索适合国内独特需求的可持续发展计划。

台北砳建筑项目的设计灵感来源于鹅卵石，建筑具有圆润优雅的外形，同时兼具力量和个性的精神。建筑以“城市客厅”为理念，立面采用玻璃和铝板，用绿植覆盖，不仅与鹅卵石上苔藓的设计概念相呼应，还提供了可呼吸的墙面，从而优化建筑的空气流通，使其成为了一个同时满足美学理念和功能需求的可持续性设计。

Aedas 与世界未来

全球视野——把世界经验引入中国

20 世纪 80 年代，中国经历了经济、政治与文化转型，伴随着新思潮和现代科学技术的推动，建筑业蓬勃发展。Aedas 率先将世界前卫的设计思想、丰富的国际经验与先进的建造技术引入中国，探索出在全球化条件下符合中国社会的设计作品，与中国共同成长。

1）关注城市空间

世界城市化背景下，Aedas 基于多元化的设计理念，对深厚的中国城市文化与变幻的城市空间进行了国际化视野下的独特解读。除关注项目本身外，Aedas 更加注重挖掘建筑与城市之间的关联性，并结合对社会与人居环境的思考，展现出城市的人文价值。

香港富临阁项目坐落于有“香港华尔街中心”之名的香港中环，以五层的亲和尺度缓和了周边密集的超高层建筑群，Aedas 在建筑中融入了丰富的城市功能，塑造出高雅自然的城市空间。

2）公共空间开放性

Aedas 强调城市公共空间的开放性与通达性，希望能够为使用者提供自发并愉悦的社交活动场所。同时，Aedas 擅长以多层次的公共空间贯连城市环境与建筑，将开放的公共空间作为建筑与交通枢纽的黏合剂，推进了公众在开放领域的自发交往，形成具有渗透性的有机建筑体。

在重庆新华书店集团公司解放碑时尚文化城设计中，Aedas 提供了 3 个不同功能不同层高的广场，贯连起休闲、餐饮、办公等一系列功能，吸引使用者汇聚的开敞空间，同时又向周围的城市街道延展，渗透到城市空间中，无形的城市精神复合有形的建筑空间，展示给重庆高密度城市核心区一幅自然的画卷。

3）建筑功能复合性

建筑的丰富性决定了感知的愉悦度。Aedas 设计了众多极具灵活性与适应性的城市综合体，也创造了许多具有丰富功能与高度连接力，富有活力、集约且可持续发展的城市枢纽。让多种功能共生以成为有机的建筑整体，Aedas 为中国提供了一种建筑功能多样化的可能性。

在新加坡星宇项目中，Aedas 保留了基地的自然地势，以阶地的形式连通城市公园与广场，一系列的坡道、自动扶梯、阶地和公共花园贯通整个综合体，形成了可穿梭的开放商场。

4）历史文化与环境特征

Aedas 重视对于当地文化特质的挖掘，试图回归文化创作的原点，着力于对建筑的个性特征的展现。在设计中，Aedas 能够巧妙结合当地自然地貌，以象征性的隐喻、图像化的造型、浪漫化的语境来表达本地传统文化。

在无锡恒隆广场项目中，Aedas 提取了中国书法的精髓，空间如同中国字画笔触之间所发挥的虚实对比一样丝丝互扣，衍生出建筑设计的动态感。公共广场包裹着历史建筑群，达到了保留历史建筑、新建建筑与新城市空间之间的丰富对话。

中国实践——将中国经验分享给全球

Aedas 从富有中国特色的城市建设中汲取经验，将在地性实践与国际化视野相结合，形成“全球与当地”的设计理念，理解并尊重中国传统文化，聚焦于中国建筑社会文化的内在关联和整体互动，凭借着丰富的国际项目经验和对城市间差异的透彻了解，建构出切合当地需要的优质环境与创新设计。

1）高层高密度城市的创新性探索

中国的城市拥有着世界独特的高密度环境，众多的高层建筑承载了大量的人类使用需求，其复杂性也是世界罕有的。Aedas 自觉投身于中国的城市建设，聚焦中国城市的经济、社会、文化背景，探索适合国内众多不同城市文化和气候的独特设计产品。

在实践中，Aedas 以敏锐的感知寻找到了“城市枢纽”的概念，以城市枢纽整合工作、生活、休闲等功能，解决土地利用的问题并深度挖掘其价值，拓展适合中国的全新建筑类型和城市设计模式。Aedas 基于在中国自身高密度城市中积累的设计经验，现正在被美国、欧洲等地区被广泛借鉴。

2）在地性与国际性的深度融合

中国拥有着深厚悠久的历史文化与丰富奇妙的地理环境，Aedas 着眼于中国城市间的差异，深入理解当地的文化、地理、历史，以在地性的思想唤回历史记忆，以象征性的设计隐喻当地文化，以有机的建筑空间呼应本地环境。

Aedas 将“全球化”的设计基础，与“当地化”的设计理念紧密结合，衍生出因地制宜的方案，更使设计项目成为改善都市环境的一种途径。现代空间和建筑体量转译传统文化中的空间和意境，在与当地的传统文化相连接的同时融合国际文化的发展变化，产生出全新的创造性设计。

3）全球化协同平台的有效搭建

Aedas 分布于世界各个地区的办公室共建起全球化平台，营造出独特的“学院式”设计氛围，通过基于协同平台的合作设计将整个 Aedas 黏合到一起，紧密相连。Aedas 在中国获得的丰富基础设施建设与综合体的项目经验等也正通过这一平台，参与到世界各地的城市建设进程中。

Aedas 与世界未来

在 2013 年中国首次提出“一带一路”倡议之前，Aedas 就已在沿线城市积累了丰富的项目经验，并将在新的城市规划策略的指导下，利用其专业知识技能和经验，为中国新型可持续城市助力，通过一带一路的拓展，进而为世界创造高效、可持续且富有活力的城市范例。

亚洲引领着全新的城市规划策略以及新型生活方式，中国也将创造全世界最具可持续性和高效的城市。通过在中国 30 余年的在地实践，Aedas 融合了国际化的理念与本土文化，将世界的经验带入中国，并通过全球平台合作共享，将在中国的实践经验运用到美国、欧洲、中东及其他亚洲国家。Aedas 将会继续积极探索创新理念与技术，与中国共同发展，参与并建构世界的未来。

Development of Aedas

Aedas was established in Hong Kong in 1985 by Keith GRIFFITHS. The name Aedas comes from the Latin aedificare, meaning "to build." Aedas recruits inspired and independently minded people who are passionate about design and want to excel; and provides them with the freedom and space to dream. Aedas was initiated with two designers in a single working space and having been in the business for 33 years. It is now one of the world's leading global architecture and design practices with preserving sensitivity of market hotspot and market requirements. According to its core strategies at different periods, it is possible to divide the development history of Aedas into four stages.

1980s – Based in Hong Kong, Focusing on High-rise and High-density Architecture

In the 1980s, the wave of urbanisation in mainland China has not yet begun and over in Europe was an economic crisis. At the time Aedas focused mainly in Hong Kong whose economic growth flourishing. Globalisation offered Hong Kong a rare opportunity to become an internationally-linked industrial and commercial metropolis, and the city quickly developed into one of the four "Asian Tiger" economies, the others being South Korea, Taiwan and Singapore.

During this period, buildings in Hong Kong were built mainly with brick and concrete. With more and more office tower and shopping center developments, there was a growing need for more complicated construction methods such as standardised and prefabricated components. From the beginning, Keith GRIFFITHS formed Aedas with an international perspective. Adopting expertise from Europe, the firm has responded with agility to architectural needs with diverse and creative designs.

1990s – Expanding to Southeast Asia with a Changing Focus on Tailor-made Architecture

In the 1990s, Aedas placed its priority on Hong Kong and Southeast Asia. Due to the economic boom in Southeast Asian countries such as Singapore, Malaysia and Indonesia, the firm gained expertise and experience in designing infrastructure, residences and commercial complexes. Aedas was thus able to expand its business to various countries and cities and provide comprehensive and professional services to clients.

The firm developed into a large corporation reaching out across Southeast Asia and the Middle East. This is the moment that Aedas shifted from assembly design to tailor-made design.

2000-2005 – Experiencing Globalisation and the Rebranding of Aedas

Between 2000 and 2005, Aedas actively took part in international competitions and paid increasing attention to the impact of globalisation. Since the 1990s, China has begun to develop complicated commercial complexes as its economy sprung up following the reforms. In the context of the worldwide financial crisis, Aedas actively participated in the urban construction in mainland China.

In 2002, Aedas opened its first mainland China office in Beijing and began to work on its first major mixed-use project in mainland China. With its rebranding and repositioning, Aedas developed rapidly and became indeed an international firm.

2005-2017: Focusing on Architectural Development in China and Paying Attention to the Local Market and Culture

Since 2005, with experience gained in rapidly developing cities such as Hong Kong, Singapore, Kuala Lumpur, Bangkok and more, Aedas has been more focused on urban developments in mainland

上海龙湖虹桥是集一站式中高端购物中心、中岛特色商业街、奢华精品酒店、办公于一体的上海西部地标型城市综合体
Longfor Hongqiao Mixed-use Project is a landmark city complex in western Shanghai, including a high-end one-stop shopping center, featured commercial street, luxury boutique hotel and offices

地铁上盖项目——重庆龙湖重庆源著综合体项目，以高度连接性极大地提升了商业价值
The high connectivity of Chongqing Longfor Yuanzhu Mixed-use Project raises its commercial value

China. It realised that internationalisation and globalisation is not enough. The diversity of China's cities prompted Aedas to deepen its understanding of local markets and the Chinese culture. It developed a global-local design philosophy to combine global expertise with local knowledge for tailored design.

Philosophy of Aedas

Aedas set different goals at different development stages. "Do what others don't, because you can do better," said Keith GRIFFITHS. By embracing diversity, innovation and a global perspective, and paying increasing attention to local social and cultural needs, Aedas brings together world-class talent and realises its vision.

Diversity, Innovation and Internationalism

Aedas creates tailor-made and innovative designs to fulfill the needs of the world's different cities. Diversity, creativity, innovation and internationalism are all part of Aedas' values.

1) Diversity

A diverse world is a more interesting world. Aedas strongly believes that a great design is a diverse design that can only be delivered by people with a deep social and cultural understanding of the communities they are designing for. It constantly looks to push the boundaries of what is possible in every new projects it undertakes.

Aedas provides services that span across architectural, art, graphic, interior, landscape, retail and urban designs as well as master planning, covering mixed-use, office, retail, residential, hospitality, infrastructure, arts and leisure, convention and exhibition, research, pharmaceutical and educational facilities.

2) Innovation

Above everything, Aedas places creativity and creative thinking. The core of Aedas lies on designers who are inspired, independently minded and passionate about design. It brings together experts from every discipline and continent to create innovative solutions. Every project, no matter how large or small, can draw on Aedas' collective wisdom from across the world to make a difference.

3) Internationalism

Aedas is the only architectural firm in the world that combines international research, local knowledge and global business. It has 12 design studios and 1,400 creative minds across the world. With its international perspective and global design standard, Aedas consciously participates in the process of urban development.

Society and Culture

Architecture connects people and culture. Aedas firmly believes in its global-local philosophy and it integrates humanity, functionality and sustainability in its design, giving concrete jungles a new and more positive definition.

Team Collaboration

1) Globalisation and Localisation

As a leading international architectural practice, Aedas has created a global platform of design excellence to share global expertise and local knowledge. This platform facilitates exchange and convergence of ideas, as well as provides resources that the designers need to create the best possible design solutions.

2) Team Collaboration

Aedas is currently managed by a 12-person Global Board. They establish and abide by strict standards and supervise the implementation. Each local office is managed by local directors and each project includes local designers in the team. Usually an Aedas project is led by two directors from different offices. They share understanding, knowledge and expertise and ensure the design is up to international standards.

3) Concept of Cooperation

Diversity, collaboration, innovation, renovation, inspiration, passion, flexibility, quality, core leadership and teamwork are the ten core concepts that sustain Aedas' global platform of collaboration. This unique structure ties all Aedas offices together and allows efficient communication.

Aedas in China

Since establishing its first office in Hong Kong in September 1985, and opening its first mainland China office in 2002 in Beijing, Aedas now has more than 30 years of experience in the architectural market of China. Its projects in China account for 60 to 65 percent of all its projects around the world. Currently Aedas has about 1,000 employees in its China offices.

"Global and Local"

1) Local Knowledge Drives Global Business

In the wave of world-wide urbanisation, one of the challenges that architectural design often faces is chaotic and "faceless" site environment. But even so there are always historical, social and cultural elements behind. Aedas employs a cultural-oriented design strategy with sustainability in mind to create socially-responsible, environmentally-responsible and historically-responsible architecture.

2) Local Practice

How to meet China's unique needs is a topic that Aedas keeps exploring. China is now at the stage with needs of continuous development. With architects who are rooted in China and understand Chinese cultural and geographical landscape well, Aedas enjoys an advantage in the local market.

Taking Xi'an Jiaotong-Liverpool University Central Building as an example, the project design drew inspirations from the famous Taihu stones which are excavated near Suzhou. The result is an elegant building that resembles a three-dimensional Suzhou garden with sustainable measures responding to local climate.

City

Globalisation has led to the rapid development of Chinese cities, but at the same time, it has diluted the awareness of Chinese architecture and culture. That has caused architecture and urban space to become generic and homogeneous. With a focus on diversity and localisation, Aedas has explored the concept of "city hubs," designing a number of world-class urban mixed-use complexes for China so that it offers an antidote to the impact of globalisation.

1) Urbanisation

In the past 20 years, around 500 million people in China have moved to urban areas. This pressure on urban land has required cities to become more intense, more efficient, and of higher density. Aedas is already exploring how to create a modern, convenient and sustainable environment for future cities. With its understanding of China's New Urbanisation Plan, Aedas believes that cities should not be expanded at the expense of agricultural land reduction. Instead, we should conduct integrated planning of existing cities and advocate a holistic and comprehensive approach to development. This means that the value of land is not simply reflected in its monetary wealth, but also its potential human values.

2) City Hubs

Aedas predicts that there will be a series of high-density city hubs surrounding the central business district of a city in the future. These city hubs are characterised by high density, high adaptability and high connectivity. City hubs will become high-density centers for living, working and recreating, with all facilities located within walking distance.

The architecture will be highly infiltrative. The ground floor will split into multiple public platforms, allowing them to be more permeable to views, light and air.

3) Mixed-use Urban Complexes

Mixed-use urban complexes are a concept originated from the West, but they can now be found in abundance throughout China. They are not only conventional mixed-use structures, but also new spatial models for social life. Urban complexes are deeply involved with every aspect of life: work, live, recreate, commute — all occuring within a short distance and creating a new urban form.

Each floor of the complexes should be fully utilized. With well-planned public space, they are well connected to transportation nodes and other developments on multiple levels, facilitating various activities and a higher quality of life. The mix of programmes leads to a mix of building forms. As conglomerates of functions, modern large-scale mixed-use projects are more flexible and adaptable.

Aedas has designed numerous mixed-use complexes in China. Evergrande Plaza in Chengdu is an example. Its design drew inspirations from the natural terraced pools in the region and the architecture reveals elements of the natural landscape. The retail component is interwoven with public space through a porous and traversable design. The project strikes a perfect balance between form and function, creating an urban oasis in the vibrant city center.

Historic Context

German author Goethe once said, "Architecture is music frozen in time." Aedas ingeniously uses architecture to create resonance with the users, reminding them the local culture, history and collective memory of the location. Olympia 66 in Dalian was inspired by the "twin carp", a symbol of wealth and abundance in traditional Chinese culture that is typically used in Chinese New Year paintings. A modern building with an Asian ethos, it responds well to the surrounding and tells a vivid architectural story to the users.

Sustainable Development

It is an architect's responsibility to create culturally and environmentally sustainable architecture. Aedas considers nature to be part of the human experience. Even people who are born and raised in crowded cities should embrace the nature. To that end, Aedas established a sustainability team, exploring ways to more sustainably developed projects in China.

The design for Lé Architecture in Taipei drew inspiration from river pebbles, developing a unique aesthetic concept that conveys the idea of roundness and elegance, as well as strength and character. The perforated green facade of the building, like mosses on pebbles, brings air into the office space. Interwoven with a series of interlocking "urban living rooms", the building is no longer a monolithic, static, impenetrable object, but a permeable, breathing mass with life.

一带一路沿线，联合利华印度尼西亚总部大楼将这一大型跨国公司的愿景、价值观及欧洲传统与当地环境和谐相融

Along the Belt and Road Aedas designed the Unilever Headquarters in Indonesia, which reflects the client's vision, values, European tradition and local context

台北砳建筑为台湾迄今最高的绿植墙办公建筑（注：统计至截稿 2018 年 1 月）
Le Architecture features Taiwan's tallest green wall for office buildings (as of January 2018)

Aedas and the Future

Global Vision: Bringing Worldwide Experience to China

China's economy, politics and culture changed drastically in the 1980s. The building industry flourished due to new ideas and modern science. Aedas brought avant-garde design ideas, ample international experience and advanced technology to China. Aedas has grown alongside with China and design work that fits into the Chinese society under the influence of globalisation.

1) Concern for Urban Space

Rather than focusing narrowly on the projects, Aedas looks into the relationship between architecture and city. By reflecting on the social and living environments, it uncovers the city's human values.

The Forum at the heart of Hong Kong's version of Wall Street, Central, is a five-storey building that stands out among the skyscrapers. Aedas integrated multiple urban functions into the building, sculpting a graceful and natural urban space.

2) Openness of Public Space

Aedas emphasises the openness and accessibility of public spaces and attempts to create pleasant public spaces that encourage social activities. Aedas is also good at connecting the buildings with multi-level public spaces, facilitating human interaction and gluing together architecture and transportation facilities.

For the Chongqing Xinhua Bookstore Group Jiefangbei Book City Mixed-Use Project, Aedas creates three distinctively themed plazas at different levels to link up various programmes within the complex, and connect the project with the surrounding streets and roads. An intangible sense of civic spirit, combined with a tangible architectural space, gives the high-density city a breathing room.

3) Multi-function

Aedas has designed many mixed-use urban complexes and has participated in the creation of highly-connected, vibrant, compact and sustainable city hubs. It presents a new multi-functional architectural possibility for China.

The Star in Singapore preserves the natural terrain of the site while connecting to nearby city parks and gardens. A series of elements including ramps, escalators, terraces and public garden allows people to traverse through.

4) History, Culture and Environment

Aedas puts enormous effort in embedding the architecture into the local history and culture. It skillfully integrates natural landscape, design metaphor and pictorial form in its designs.

When designing Center 66 project in Wuxi, Aedas drew inspiration from Chinese calligraphy. Fluid and energetic strokes form the volumes and spaces, like the play on positive and negative spaces in a Chinese painting. The preserved historical buildings sit at the new urban plaza, creating a dialogue between the old and new.

China's Practice: Sharing the China Experience with the World

Aedas has gained much experience in China. Its global-local design philosophy enables the delivery of innovative designs that are tailored to the needs of the local cities and communities.

1) Innovative Exploration of High-rise and High-density Cities

Chinese cities are uniquely high in density. Numerous high-rise buildings have been built to house a large population, creating complex urban environments that are rarely seen in other parts of the world. Through its work in China, Aedas sensibly comes up with the "city hubs" concept and advocates the live-work-recreate dynamics; and as a result, ease the pressure on land use while conceiving a new building typology and model for urban design. The high-density China model is now being referenced to in USA, Europe and other parts of the world.

2) Integration of Global and Local

China is home to a profound cultural history and diverse geographical features. Aedas calls on its global expertise and local knowledge to create architectural spaces that reflect the local context, history, culture and geography.

3) Global Platform for Collaboration

Aedas' global platform allows its experience in designing large-scale mixed-use and infrastructure facilities in China to be shared on a global level, impacting urbanisation in other parts of the world.

Aedas and the Future

Before China introduced the Belt and Road Initiative in 2013, Aedas was already designing for cities along its land and sea routes. Under the new urban planning policy, it will continue to contribute its professional knowledge and experience to the development of sustainable cities in China and along the Belt and Road.

Asia is leading new urban planning strategies and new lifestyles, and China is about to build the most sustainable and efficient cities. Over the past 30 years, Aedas applied global expertise and local knowledge in China; and through its global platform shared such experience with other Asian countries and the western world. Aedas will continue to develop with China and together we contribute to the future of the world.

Aedas Arts Team 近期完成了位于伦敦西区中心位置的维多利亚皇宫剧院的翻新工程
Aedas Arts Team recently completed refurbishment of the century-old Grade II Listed Victoria Palace Theatre building in the heart of London's West End

九项思考 Q & A
Q&A on Design Trends

文 徐洁 by XU Jie

一、《国家新型城镇化规划（2014—2020 年）》

中国正处于高密度城镇化发展阶段，Aedas 对城镇化持有怎样的态度？

“城镇化”是伴随工业化发展、非农产业在城镇集聚、农村人口向城镇集中的自然历史过程，也是国家现代化的重要标志。城镇化的进程并非为中国独有，全球不同的国家、城市都在用自己的方式利用土地。2014 年 3 月，中国国务院颁布了《国家新型城镇化规划（2014—2020 年）》（以下简称“《规划》”），对开发紧凑型城市提出了一系列目标，从而开启了中国城镇化发展的新纪元。

Aedas 认为中国城镇化策略是极大的进步。在《规划》中，土地不单单以其财富价值来度量，同时也是当地社会人文价值的空间载体。在此基础上能清晰地理解到中国城镇规划的丰富内涵，例如：拒绝以侵吞自然耕地为代价的盲目扩张，整体规划既有城市，向全局化、精细化发展；夯实城市基础设施，并借助大数据与智慧城市的前沿理论技术分析、整理；尊重每个区域的历史文脉，并将其融入新的建设中……这些做法都是将“人与自然”的发展纳入城市发展的考量范围中，在提升生活品质的同时推动城市的可持续发展。这是 Aedas 理解的新型城镇化。

北京新浪总部大楼是在美国上市、拥有网站和微博的新浪在中国的主要办公楼
Sina Plaza is a headquarters development located at the Zhongguancun Science Park in northwest Beijing. It is the principal office of Sina Corporation in China

具体来说，Aedas 将怎样应对中国城镇化的趋势？

Aedas 看好中国城镇化，拥抱这一发展趋势，通过我们的设计提升人们的生活品质。我们要重新设计人们在城市中发生的行为，不仅是居住、娱乐、工作。

举例来说，《规划》中指出，“使交通距离的最小化和开发过程中对现有文化、社区的包容性”。作为设计者，我们可以缩短有效的步行距离，营造出不同场景，使行走过程富有趣味，人们可以路过有趣的店铺和景观，获得更多可以享受、彼此交流的公共空间；在设计商业综合体时，我们会考虑公共空间可能具备的更多功能，我们会架设天桥来提高交通可达性，它可以通往咖啡馆、办公楼大堂、公园绿地、电影院的售票中心，通过这样的方式为当地居民带来便捷，带来积极而潜移默化的生活改善。Aedas 认为，“互动”可以为城市创造更多价值，使之回到城市最根本、最原始的初衷，即让人与人产生聚集与交流。因此，我们强调包容性，这既包括对文化的包容，也包括对人群的包容。我们以这样积极的心态来迎接中国城镇化的建设发展。

徐洁 | XU Jie

同济大学建筑与城市规划学院副教授，《时代建筑》杂志执行主编，中国建筑学会建筑传媒学术委员会理事。

Associate Professor of the College of Architecture and Urban Planning, Tongji University; Executive Editor-in-chief of *TIME + ARCHITECTURE* Journal; Council Member of Media and Communication Committee, Architecture Society of China.

二、高密度城市枢纽

请简要解释“高密度城市枢纽”。

城市的发展一直伴随着社会经济的发展，也催生出不同的建筑类型。尽管如电梯、地铁、高速公路等交通技术的发展为建筑高度不断突破、城市密度增大带来了可能，然而城市的飞速发展是为了将人们带往更高效的生活方式。我们观察到，不同于以往功能简单的商办项目，近年来中国开始涌现出大量复合功能的综合体项目。这些项目整合了周边环境，是全面涵盖功能空间的高密度综合体。产生这一现象的大背景是中国城市将“生活—工作—娱乐”紧密相连。城市中对于超高密度的城市中心开发已经基本完善，现在进行中的是开发城市次一级的功能综合中心，我们称之为“城市枢纽（City Hubs）”。这些高密度的城市枢纽围绕着传统的中央商务区，以极便捷的交通条件与城际高铁系统形成良好的连接，它们在北京、上海、深圳方兴未艾。不同于西方传统的中央商务区以单一的办公空间为主导，高密度城市枢纽拥有更加丰富、灵活的城市社会生活方式。我们也许会将其称为“中央商务区 +”，或者“中央文化商务区（Central Culture Business District，简称‘CCBD’）”，即文化（cultural）、连接（connected）、便捷（convenient）。

是 Aedas 发明了“城市枢纽”一词吗？其优势体现在哪些方面？

Aedas 没有“发明”这个词，我们只是准确地感知、认识到它，并且对其进行拓展。

土地利用的压力要求城市紧凑、集约、高密度，而城市枢纽正是城市规划发展中解决该问题的集大成者。城市枢纽将成为高密度的“工作、生活及休闲中心”，所有设施都在步行可达的范围内。每一处城市枢纽处于地理位置极佳的连接节点上（这一点我们可以通过地铁网的密度来预测未来城市枢纽的位置），通过整合各功能，人们能够在这个空间中会面交流，以此开启高连接性的城市中心。

在上海，居住区、交通设施与大型公园紧密结合，形成了良好的生活环境。一方面，我们需要建设更多的城市次中心和工作空间，为城市居民提供更多交流的可能性，降低每日通勤时间，提高城市的活力、效率和城市连通性；另一方面，在枢纽之间的区域可以转化为绿地和公园，成为较低密度的空间。通过引入城市枢纽，可以为城市居民提供一个现代、便捷、通达且可持续的丰富环境。

我们在重庆解放碑区建设中的新华书店项目地处城市核心区，这家书店所承载的不是单一的书籍交易职能，而是集合了阅读、餐饮、娱乐、办公、会晤等多种功能于一体的空间，也是一座体现重庆特色，能够与解放碑商圈形成互动关系的商业综合体，其超强的商业凝聚力与交通便利度使项目成为独树一帜的新型地标，它也是城市枢纽的一种模式。

三、具有通透性和连接性的设计

Aedas 为什么在设计理念中强调通透性和连接性？

首先，我们注意到，近年来中国出现了很多新型的建筑形式，包括 LOFT、商业住宅、商业零售村、SOHO 等，单单依靠地面层公共空间的连接已经无法满足高密度城市的发展需要，所以我们开始设计了竖向多层次的公共空间和交通连接；其次，社交媒体的迅速发展曾经一度让人们以为不需要经常会面，网络可以解决一切沟通，但人们终究对远距离的线上交流感到不满足和厌倦，内心亟待真切面对面的交流。因此，Aedas 希望为人们的相聚和交流创造空间，塑造积极、开放的城市性格，从而让人与人之间的交往变得更加简单。

Aedas 将采用哪些具体方式增强建筑间的通透性和连接性？

自从认识到精雕细刻的“高层加裙楼”的形式已经过时，不再能满足中国新的发展后，Aedas 就开始通过多样化的方式增强建筑间的通透性和连接性。不同于仅仅只在地面一层多开设几扇门让不同的建筑体之间有渠道相通，我们希望用多种方式创造多层天桥、天台、空中花园和地下室等公共空间的连接，并且渗透到整个项目中。Aedas 的设计摒弃了单一塔楼的旧概念，采用更加具有连贯性的城市模型——在垂直和水平方向双向连通、更具适应性和灵活性的建筑形式。例如，我们为苏州西交利物浦大学设计的中心楼就很好地体现了这一点：这座楼是集行政办公、图书馆藏、信息中心于一体的综合项目，我们通过增加地面一层和二层的入口、六层的灰空间庭院、九层的连廊以及十二层的瞭望平台等，大大增加了公共空间的活力与人群停驻交流的可能性，在丰富空间的同时，也让建筑形体变得轻盈通透，富有活力。

四、新型商业零售

Aedas 怎样看待新型商业零售的趋势？

无论在建筑形式或建筑功能上，今天的商业零售无疑正在发生巨大的转变。一方面，零售的意义更加多元，从单一的消费场所转变为兼具社交、娱乐、餐饮等多功能的场所；另一方面，新零售的市场越来越细分，更加有针对性，产生了分别针对高端人士、家庭、时尚达人等不同市场人群的购物中心。中国零售市场的空间职能发生的变化也与百姓的社交方式有关。在西方国家，人们乐于邀请朋友们到家中做客，因为他们有更多的居家空间去接待友人；而当下中国的城市居民更乐于与朋友们相聚在休闲消费场所，一起喝杯茶、吃个晚饭，于是购物中心成为了一个可以让亲朋好友们会面的公共空间，可以开展丰富的活动，而“购物”只是其中一项。

网络购物的兴起也取代了大量传统单一的实体购物场所。同时，城市又在数据化进程中向智慧型、密集型转变，购物中心已成为各种城市功能的“黏合剂”，将各种业态连结，形成贯穿城市社会生活的“经脉”。可以预见，多元化的、作为社交枢纽的购物中心将成为未来发展的基本节点，这让我们清楚地知道应该为未来的商业建筑融入哪些功能：办公、酒店，甚至是住宅、服务式公寓、交通枢纽，等等。

新零售项目格外重视设计前期的调研，Aedas 在这方面有什么特殊优势？

前期调研分为正式和非正式两种情况。正式的市场调研由我们选定的市场调研公司进行，从市场专业的角度进行问卷设置并逐步获取结论；但 Aedas 也非常重视非正式情况下的调研，例如采访出租车司机，在日常生活中他们能够遇到当地人，也能遇到游客。我们通常不会直接询问客户或者和项目有关的人，因为他们会根据自己的项目产生主观的判断。我们会进入项目基地和周边，询问用餐的客人或者服务生，也许他们是本地人，甚至好几代都生长于斯，也许他们刚刚举家迁徙来此，但他们会根据实际体验告诉我们周末他们会去哪儿，喜欢怎样的活动……我们总会得到许多不同的答案，但是某些部分一定是一致的、准确的，这些重叠的内容就可以帮助我们对建筑物进行构想。

Aedas 的目标是为人而设计，我们相信，如果不与终端用户交谈就无法给出最佳的解决方案。在做基地考察时，每当我们拿出相机拍照，总会有周围的居民们围过来询问情况，这时我们会与他们交谈，因为邻里们非常关心周围将发生的变化，他们会提出愿景和建议，Aedas 通过这种方式来了解他们需要怎样的建筑来使家园变得更合理、更美好。

五、基础设施设计

Aedas 怎样看待中国城市发展过程中基础设施的建设？

凭借 Aedas 在全球轨道、港口和机场等基础设施项目上的经验，我们对中国交通网络上的巨大发展已拥有深入的了解。连接中国所有大城市的“四横四纵”一体化高速铁路网络已经临近竣工，同时中国又不断开发新机场来服务其快速发展的城市。Aedas 相继为这些新的交通基础设施设计了最为先进的建筑，例如香港和西安的轨道项目设计，以及深圳和香港的机场项目设计。现今中国的城市通过高速铁路和航空连结，乡镇通过郊区铁路连结，城市内部通过地铁系统连结。中国拥有超过全球其他国家和地区总和的高铁及地铁系统。公共交通和电动汽车的发展将很快大幅降低中国的能源消耗和环境污染。密集的城市地铁系统将使城市步行和骑行更加便捷，而这些都将促使更多的公共连接空间和通透性建筑的出现。Aedas 的设计正在积极参与到这样的未来建设中。

香港国际机场中场客运廊自 2016 年开业以来，一直深受人们欢迎
The Hong Kong International Airport Midfield Concourse has been well received by the public since its opening in 2016

人们对于机场、铁路、地铁等基础设施有一贯的旧有印象，Aedas 的设计是怎样与同类设计公司相区别的？

机场是城市的门户，代表了这个地区和它的文化，需呈现独特的地区性格，这是 Aedas 热爱航空设施的原因——它具有无界性。随着机场和空间的更新发展，商业零售业态已成为机场巨大的收入来源，免税店购物成为旅客们迫不及待的活动行程。另外，瑜伽教室、泳池、电影院、博物馆、画廊都存在于全球的机场中。

人们对于旅行的观念已经改变，旅客对机场能提供的服务的期待也变得巨大。我们认为这是一个从未有过的全新时代，酒店、商业、休闲、交通以如此整体的方式紧密结合。休闲活动的兴起使得旅客的期待越来越高，他们期待机场可以成为目的地之一，希望可以尽快开启机场的丰富体验。中国是如此的巨大和多样，文化、食物、人文、气候都非常丰富，各个城市又是各有特点，在中国各地进行机场设计具有本质上的不同挑战。“十三五”对于中国航空事业的发展有着迅猛的促进作用，这令世界瞩目，这一愿景也时刻激励着 Aedas 继续前进。

六、中国的特大城市与城市设计

针对北京、上海、广州三座城市，其城市发展已经基本完善。Aedas 怎样看待其未来的发展空间？

中国已有三个全世界规模最大的大型城市群——珠江三角洲（12 000 万人口）、长江三角洲（8 800 万人口）及环渤海经济圈（6 600 万人口）。我们开始看到中国，尤其是内陆城市经济复苏的趋势，并承接了多个城市重新规划的委任，这与我们在 2012—2013 年间为新城市所做的规划很不一样。这些城市的重新规划，或可被称为城市更新，是发达城市的主要趋势。当然，这一趋势与诸如城市枢纽等其他要素互相关联。中国已经着手创建世界最高密度的智能城市，以适应现代互联网生活—工作社会的需求。以上海为例，如今其住宅和商务区大多是分离的，通过在现有基础设施节点上建造城市枢纽，可以在高价值土地上开发高密度生活—工作社区，建立良好的连接性和效率，并能更好地开发利用城市中现有的老旧项目。

针对另一批发展中的城市群，例如“香港—深圳—广州”“上海—杭州—宁波”“重庆—成都”“石家庄—郑州”等，Aedas 有怎样的期待？是否有建设战略部署？

Aedas 是将“全球化”和“当地化”相结合的设计集团，在任何一座进行设计服务的城市都不将我们视为外来公司。为了了解当地文化、气候和经济，我们在所服务的每一个区域都设有办公室。我们位于北京、上海、成都、深圳和香港的办公室，都深入了解各自所处的中国北部、东部、西部和南部的情况，并将这些知识分别传递给其他在华办公室的设计师们。重要的是，出于对中国国情特殊需求的理解，以及这 5 个办公室所拥有的高层与高密度设计的丰富经验，Aedas 所有在中国的设计均由这 5 个办公室操刀主持。我们将基于这 5 个办公室，继续拓展从事其他发展中地区的项目，并在需要时将办公网络扩展到更多中国城市。

七、文化和故事

“当地化”是 Aedas 设计项目的重点之一，你们是怎样将当地文化恰如其分地注入到设计中的？

Aedas 在中国拥有 5 个办公室约千名员工，对中国有着深入的理解。在 Aedas 的项目中，我们创建能帮助理解当地文化、地理、历史和经济的所有条件，并将获得的理解融入到设计中，使我们的设计具有相关性和影响力。我们特别注重将当地文化尤其注重将历史和传说融入设计，这使得我们的每一个设计能贴合所在城市的地域和文化特色。更加重要的是，我们利用建筑设计追溯对于当地文化、历史事件和地理环境的回忆，能使项目与城市紧密依存，唤起人们记忆中精彩有趣的故事并代代相传。

Aedas 尊重历史文脉的延续。在无锡恒隆广场，项目基地中央有一个两侧配有戏台的双层戏院以及一个始建于 14 世纪的明朝城隍庙古戏台。设计将其保留下来，并让老建筑主导了整个地块新建筑的规划，同时串联起钱钟书故居和现存的清泉古井。我们将大型公共景观广场与既有保护建筑整合到一起，形成连通一体的有机建筑群，促进历史建筑与新城市空间之间的对话。

在述说故事的时候，Aedas 在设计中植入了什么新的元素？

建筑自身所承载的历史，是一个非常好的故事起点。但对新建建筑而言，我们需要找到参照，通过一些比喻或象征来进行表达。客户总是很容易理解“功能”，但是建筑的“形式”关乎美感，通常就比较抽象而主观了，因为这更多地是与造型、颜色或材料感性地关联在一起。因此，Aedas 会利用一些隐喻来帮助人们理解某种形式，帮助他们去接受一座建筑。

以大连恒隆广场为例，我们在外观形式满足功能要求的基础上，赋予其“如意双鲤”的概念，立即获得了当地政府领导和业主的理解和欣赏。2008—2009 年那两年，建筑颇为流行以具象的概念进入公共视野，例如“水立方”“鸟巢”等。建筑有这样一个好的寓意、好的标志能让大家都记得认可，能汇聚更多的城市活力要素，并在广泛流传中更长久地传承下去。

八、一带一路

请谈谈一带一路，特别是Aedas对“海上丝绸之路”的看法，并系统、具体地说明 Aedas 在城市规划、项目建设等方面是如何呼应一带一路的倡议？

海上丝绸之路从中国南部发端，跨越东南亚和非洲，经海路一直抵达欧洲。在海上丝绸之路上，Aedas 与中国开发商在沿线的合作由来已久，比如绿地集团位于马来西亚的开发项目，招商地产位于非洲吉布提的港口等。香港富临阁将一座公园架高于地面之上，并与中央商务区的其他项目相连，这个项目已成为多层公共空间的一个范本；香港国际机场中场客运廊和北卫星客运廊提供了高度灵活，且适应未来需求的现代旅客体验。在新加坡，星宇项目打破了热带国家须建造有空调系统的购物中心的“规则”，且连同 Sandcrawler 项目一起为当地扩展了公共空间。我们设计的联合利华印度尼西亚总部大楼，其设计成功地将这一跨国公司的愿景、价值观及欧洲传统与印度尼西亚当地环境和谐相容。在迪拜，迪拜地铁站是中东地区首个地铁节点，并成为该区未来铁路发展的标准。吉布提港口重建总体规划通过重新安置并扩展港口设施，重建吉布提旧港，吸引外资投资该城市并扩展至整个国家。在伊斯干达，高密度生活—工作原则被应用于马来西亚新城，这完全基于我们在中国的经验。在伦敦，摄政区（Regent Quarter）重建项目为该区带来了活力和经济增长。

Aedas 是否会将中国的经验带给世界？

中国城市具有高密度的环境，大量人群居住在高层建筑之中，这种奇妙的环境所孕育的设计经验是世界少有的。中国最初追随了美国的规划与商业楼宇开发模式以及香港的商业综合体开发模式。但到 2008 年，中国已在开发适合不同城市文化和气候的独特设计产品，实践出全新的建筑类型和城市设计模式。

阿联酋迪拜地铁是全球最大规模、最先进的自动交通项目之一，全长 74km，共有 45 个车站，其中包括 9 个地下车站和 2 个转运站以及操作控制中心
Dubai Metro is one of the world's largest and most advanced automated passenger transit projects with 74km of rail and 45 stations, 9 of which are underground stations and 2 are interchange stations and control centers

相当数量的西方建筑师，并不理解高密度与高层建筑。许多纽约与伦敦的建筑师，对一栋建筑需要多少升降电梯没有经验，缺少对结构的敏感度，因为在纽约要重建高层建筑是十分困难的。伦敦的塔楼事实上在造价高昂的同时效率较低，需要花费大量的时间建造，现实中它们相较于建筑功能体更具有城市雕塑的存在意义。在伦敦，220 个高层塔楼分散在城市的各个地方，而伦敦并不是高密度城市，基础设施网十分薄弱，无法承载众多分散枢纽间的交通，所以伦敦需要考虑的是如何在城市中分布这些塔楼和高密度社会生活。而在中国的很多城市，高密度被集合起来，从而发挥出更高的利用率和城市活力，这让我们想到了上海地铁线路的高密度发展带给人们更加便利的生活，通过研究地铁轨道线路和上海的城市枢纽发展，我们可以提前预测地块的价值。当看到两至三个地铁站点集合在一起时，就可判断这块地十分宝贵，将成为城市枢纽。中国的经验可以传播给世界各地的城市，促进其发展。同样，在制造业方面，中国的经验也能分享给世界，这源于中国极大的内需市场。比如办公高层建筑，其核心筒的大小会影响到实际办公区域的空间质量，因此在设计上需要尽可能缩小核心筒、使电梯更高效。电梯生产商已经在制造双轿厢电梯，甚至能与智能手机控制相联系。有很多设备的系统开发都是针对中国市场研究展开的，已在全球处于领先水平，这些技术与经验非常值得在全球其他城市推广。

九、城市改造与修复

Aedas 在城市改造项目中有怎样的思考？

Aedas 对当地情况的深刻理解反映在我们针对项目进行考察，并向当地政府和土地所有者提供改造与改善的建议中。以香港的中央商务区为例，Aedas 推动香港中央商务区的设计已有逾 30 年的经验，项目包括兰桂坊娱乐区和香港置地商业零售及办公中心等。Aedas 在被称为“香港版华尔街”的中环，活化、翻新、重建了一系列主要建筑，并将其转变为一个全天候、多层次的“工作—生活—休闲”城市中心。我们对市场需求、当地文化及其变化了如指掌，善于将本身只具有单一银行办公功能的区域打造成为集商业零售、酒店、办公楼和休闲娱乐为一体的、世界上最成功的中央商务区之一。作为建筑师，我们通过这样的方式来优化都市环境。

再以历史剧院为例。Aedas 在 20 世纪开始修复英国历史剧院，当时政府与民众对于建筑保护和保护措施的态度与现在是截然不同的。在维多利亚和爱德华七世时期（1860—1915 年间），剧院在公众规划部门和建筑社区中的评价很低，民众对于剧院普遍缺乏关注，且任何关于历史剧院的修复都需要规划部门予以开创性的审批许可。Aedas 的作品诺丁汉皇家剧院始建于 1865 年，翻新开始于 1978 年。在非专业人士的眼中，剧院看似保存完好，然而实际上建筑改造及后期干预的工作量巨大。历史表演空间和公共空间的建筑处理方式差别巨大：公共空间使用了更为现代的设计语言，表演空间的历史获得充分尊重。40 年来，随着全球对于历史建筑价值的不断关注，还有观众态度的转变，我们对待和改造历史剧院的理念和技术也变化显著。随着留存的历史剧院越来越少，且一些仍然处于危险之中，它们作为一种建筑形态的固有价值在增加。如今的工作有着更为严格的规定，需要与当地以及国家历史保护组织紧密合作。

Aedas 的决策工具，是适用于每座建筑的详细保护清单，以此分析历史价值和剧院发展可行性，评估其社会重要性和文化价值，并决策是否采取加建或拆除措施。作为设计师，需要对重建方案做出更明智的细节决定，需要更加理解历史材料和施工技术。因此，当我们平衡历史元素、现代娱乐以及观众期待时，有能力做出专业判断。无可争议的是，正如同其他保护领域的设计师，我们对于历史剧院的设计与修复的心得、经验及专业知识，在过去的 40 年间不断在进步。

关于城市修复，特别是历史建筑的专项修复工作，请谈一谈 Aedas 持有哪种态度？

要视情况而定。以无锡恒隆广场为例，我们保留了城隍庙区，并将其作为该项目大型公共广场的核心。该历史建筑群包括始建于明朝（约 1369 年）的内戏台、外戏台和西偏厅 3 座古建筑。原有的两层高戏台按原貌修葺后已恢复昔日的辉煌。围绕着历史建筑的公共广场强调对建筑物的历史共鸣，同时创造出与现代环境的强烈对比。

另一个例子是位于香港的艺术社区——湾仔茂萝街 / 巴路士街活化及文物保育项目。该项目复兴了一组稀有而完整的市区历史建筑群，令其成为兼具公共休闲、文化及创意工业的现代场所。一些元素以现代技术进行了修复，而一些则是全新的。通过采用创新的设计，我们重新诠释了空间运用和材料使用的传统智慧，并改进原结构及设施使其符合现代建筑规范与需求，延续都市传统文脉的建筑使命。

我们还在位于英国伦敦的摄政地区对已有的建筑进行适量的功能更新和修复，又在周边规划穿插进全新的建筑，通过一系列的策略来整体振兴该地区。英国在决定建造新的建筑之前，艺术领域的修复和翻新是经过深思熟虑和成本估算的。如果一座建筑需要翻新、重建或者替换，那么广泛的公共协商会确保项目不仅仅是增加土地价值，还能增加公共使用性以最大化社会价值。Aedas 所有修复项目和新建项目的底线被称为“四重底线”：项目的社会、文化、经济和可持续发展价值的最大化。

以位于英国唐卡斯特文化社区的 Cast 剧院为例，不同的剧院和客户有不同的需求；建筑可以被永久修复，或者通过天衣无缝的改造和加建成为一个空间序列，又或者在他们的想象中更为前卫。在 Aedas 的作品中，把这些不同的处理方式连接起来的是共享价值，即历史建筑元素如何启发建造剧院的创造过程——让这些承载着共同文化遗产的建筑重回我们的社区，令市民为此倍感骄傲。

中国和英国类似，有许多活跃的机构致力于历史建筑和建筑遗产的保护。在英国，这些机构上至国家英格兰历史和剧院信托团体，下至当地政府层面的地方保护部门。此外，还有很多其他关注群体，例如 20 世纪协会、维多利亚协会等，都致力于历史保护。在英国，修复一座历史建筑在工作开始之前需要获得这些机构的许可。英国至今已经实行了一段时间的 20 世纪建筑保护，但中国从 2016 年的首部 20 世纪建筑遗产名录公布才刚开始着手实施对于历史建筑和建筑遗产的保护。两国可以相互学习的地方非常多，在历史建筑修复方面，特别是在专业工艺技能方面。随着预制建筑的推广，石匠、木雕工、泥瓦匠这些工匠的技能在减退，修复这些历史建筑的过程也是保护这些技能的过程。

1. *National New Urbanisation Plan (2014-2020)*

China is currently in the phase of high-density urbanisation. What is Aedas' take on this matter?

Urbanisation is a historic and natural process that stems from industrial development, the urban aggregation of non-agricultural industries, and the migration of rural population to urban areas. It is also an important sign of national modernisation. Urbanisation does not only occur in China; different countries and cities worldwide have developed distinctive forms of land use. In March 2014, the Chinese Government has published the *National New Urbanisation Plan (2014-2020)* — known as the Plan for short in the following text — which presented a series of goals for developing compact cities. The Plan opened a new era for the development of Chinese urbanisation.

Aedas regards the Plan as a great step forward. It measures land not only by its property value, but also as a spatial carrier of local social and human values. With this way of thinking, we can clearly understand the richness of the Plan's goals. For instance, plan cities comprehensively to prevent blind urban sprawl at the cost of wasting farmland; firm up urban infrastructure through big data analysis and smart city initiatives; and respect the historic context of every region and incorporate it into new construction. By integrating the built and natural environments, the Plan promotes the sustainable development of cities, while enhancing people's quality of life. This is what Aedas understands as the new paradigm of urbanisation in China.

More specifically, how will Aedas deal with the trend of urbanisation in China?

Aedas is optimistic about Chinese urbanisation. Embracing the current trend, our designs promote an emphasis on quality of life. We would like to redesign human behaviour in the city, not just in residential area but also in entertainment spaces and workplaces; for instance, creating stimulating pedestrian areas with efficient connections, attractive shops and enjoyable landscapes, so that people can enjoy a joyful social life.

The Plan mentions "minimising transport journeys and enhancing the inclusivity of existing cultures and communities." As designers, we may shorten the effective walking distance while creating various environments that enrich the walking experience. Pedestrians may walk past exciting shops and scenes while making their way to public spaces where they can relax, gather and interact. When designing a commercial complex, we consider how the public space may serve multiple functions. We could propose a sky bridge to improve transport accessibility. The bridge could lead to different places such as cafés, office lobbies, parks or cinemas. These would facilitate local residents' daily lives while actively enhancing their quality of life. Aedas believes that public social interaction creates value for the city and its original purpose, which is to facilitate human gathering and interaction. We value inclusivity, not only on a cultural level, but on a basic human level. We greet the development of Chinese urbanisation with a positive attitude.

2. High-density City Hubs

How would you explain your concept of high-density city hubs?

Urban development is always accompanied by economic development, and it also leads to the birth of distinctive architectural forms. With new, more efficient transport technologies, whether it concerns elevators, subway systems or highways, there are more possibilities for taller buildings and even more dense cities. Urban development requires the same efficiency. Rather than office-commercial projects with simple functions, large mixed-use projects have been emerging in China in recent years. These projects integrate themselves into the surrounding environment, forming a high-density complex that contains all functions. China's unique way of blending life and work is the driving force behind these high-density mixed-use developments.

The development of ultra-high-density hubs at the heart of cities are mostly completed. Now Chinese cities are developing secondary mixed-use city centers of very high density, which we call city hubs. These surround the traditional CBD and are already emerging in Shanghai, Beijing and Shenzhen. These city hubs are rich in transport connectivity and are well linked to the surrounding urban area as well as to the national high speed rail system. They have a rich and vibrant life not seen in traditional Western CBDs, which are generally limited to office use. So we could call them CBD+, or maybe CCBD (Central Cultural Business District), for Cultural, Connected and Convenient.

Does Aedas invent the terminology "city hubs"? What are the advantages of city hubs?

Aedas did not invent this terminology. We saw it, recognised it and then expanded its meaning.

Population pressure requires cities with high density and intensity. City hubs are the right solution to such urban planning challenges. They become high-density centers for living, working and recreation, with all facilities located within walking distance of each other. Every location of the city hubs can be considered depending on its geographical connections, or depending on the density of the subway network, for instance. By integrating multiple features, the highly connective city hubs will successfully allow people to meet within a certain space.

In Shanghai, residential areas, transportation facilities and parks are tightly connected, creating a fairly high quality living environment. On one hand, we require more construction of city hubs and working spaces to gain more opportunities for city residents to interact, while improving their livelihoods and decreasing the hours they spend commuting. On the other hand, the space between each hub is conceived as parks and green areas, which will be lower density. By introducing the city hubs, residents can enjoy a new way of living that is more convenient, inclusive and sustainable.

Our Xinhua Bookstore Group Jiefangbei Book City Mixed-Use Project, which is now under construction, encompasses a cultural plaza, retail, apartments, offices and a boutique hotel in the Jiefangbei Central Business District, the heart of downtown Chongqing. This multi-feature space embodies Chongqing's unique characteristics and also interacts with Jiefangbei District as a commercial complex. This gigantic development creates an exceedingly strong cohesive force that also forms a brand new civic landmark that flies its own colours.

3. Infiltrative and Connected Design

Why does Aedas emphasise infiltrative and connected design?

We notice the emergence of new building types in China: LOFT, serviced apartments, village retail, SOHO. Simply utilising the ground floor as public space and connectivity cannot fulfil the development requirements of high-density cities. So, we have started to create new public space and connectivity at many levels to better "irrigate" our urban environment. Also, the vast growth of social network services has disconnected people from having face-to-face interaction, and many people only communicate through the internet. But this is unsatisfying and alienating, and people feel an urge to communicate face-to-face. So Aedas would like to create spaces where people can gather around and interact. A positive and accessible city can bring simplicity back to human connection.

What specific methods will Aedas employ in order to strengthen infiltrative and connected buildings?

The previous "podium and tower" development model is outdated and irrelevant to the new direction of China's development. The density is such that public space and connectivity at ground level is insufficient, so Aedas designs public space and connections at many levels. Aedas' designs ensure developments are a series of infitrative and connected buildings linked by multi-level bridges, sky decks, sky parks and basements, which infuse the development with public spaces. The firm's designs dematerialise the old concept of singular towers into a more cohesive urban form of adaptable and flexible building plates connected vertically and horizontally. For instance, our design for Xi'an Jiaotong-Liverpool University's new Central Building has a good approach to connectivity. The building is designed as the combination of administrative offices, library and information center. The design widens the possibility of interaction and vitality in the public space by adding entrances on the second and the ground floors, a courtyard on the sixth floor, a corridor on the ninth floor and an observatory platform on the 12th floor. These spaces allow the architectural form to be light and transparent while also infused with energy.

4. New Retail

What is Aedas' take on the changing retail environment?

Retail developments are changing drastically in both type and use. On one hand, the retail scene is more diversified, developing from a place for shopping to a place for social activities. On the other hand, retail becomes more and more specialised. There are diffesent shopping centers targeted to customers such as the wealthy, families, fashionistas and so on respectively. The biggest change is that retail now relates primarily to lifestyle. In Western countries, people like to invite friends to their home since they have more living space to host a gathering, whereas Chinese people are more willing to invite their friends to leisure spaces for a cup of tea or a meal. Therefore, the shopping center becomes a social place for people to meet their family and friends, with shopping being just one of many different activities that take place there.

The boom in online shopping is also taking the place of the traditional offline shopping center. With cities adopting more digital infrastructure, becoming more compact and more "smart," the shopping center becomes the "glue" that holds together the city's various functions. That means shopping centers will inevitably become an essential element of future development. This makes us aware of the functions that should be integrated into a shopping center: workplaces, hotels, residences, serviced apartments and transport hubs.

New retail projects pay special attention to pre-design research. What special advantage does Aedas have in this regard?

Pre-design research can be categorised into formal and informal. The formal market investigation is conducted by a selected market research firm, which draws its conclusion based on a questionnaire survey informed by a professional view of the market. But Aedas places emphasis on informal investigation as well, such as interviewing taxi drivers who can speak to the needs of local people, as well as tourists from far away. We do not directly ask consultants or other such individuals due to their subjective opinions. We normally approach a project's site and surroundings, talking with customers of nearby businesses or waiters in the local restaurants, even residents, whether their families have lived there for centuries or they have recently settled there. They will share details of their weekend getaways, the activities they enjoy. We always receive different answers, and the ones that share something in common influence our design.

Aedas' mission is to design for human beings. We believe it's impossible to find the best design solution without communicating with the future users of a building. When taking photographs during site research, there are always curious residents who surround us. We take the opportunity to communicate with them since they know a lot about the changes in their community. They will provide their visions and suggestions. By talking to them, Aedas can understand what they need from the building we are designing in their community.

5. Infrastructure Design

How does Aedas view infrastructure in the development of Chinese cities?

Through Aedas' experience in infrastructure such as rail, port and airport projects, we understand the great advances China has made in coordinating its transportation networks. The "four vertical and four horizontal" integrated high-speed rail network connecting all the great cities of China is nearly complete. Meanwhile, China continues to develop new airports to serve its rapidly increasing urban population. Aedas designs state of the art terminal buildings for these new transportation infrastructure. Aedas is designing rail projects in Hong Kong and Xi'an, airport projects in Shenzhen and Hong Kong. Chinese cities are now connected by air and high-speed rail; towns are connected by suburban rail; and within cities there are subway systems. China now has more high-speed rail and subway lines than the rest of the entire world combined. Public transportation and the electric car will soon massively reduce energy consumption and pollution in China. The dense city subway system will facilitate walking and cycling within the city and this will lead to a drive for even more connected public space and infiltrative buildings. Aedas is designing for this future.

People have an outdated perception of infrastructure such as airports, railway, and subway systems. How does Aedas distinguish itself from the other companies in designing infrastructure?

An airport is the gateway to a city and represents the region and its culture. It needs to convey a region's unique character. This is why Aedas loves aviation facilities — they are unbounded. With the use of materials and space, commercial retailing has become a huge source of revenue for the airport. Duty-free shopping has also become an irresistible draw for tourists. Yoga classes, swimming pools, cinemas, museums and galleries all exist in airports around the world.
The concept of traveling has changed; so have the expectations of tourists. We think this is an unprecedented era. Hotel, retail, leisure, transportation and all sorts of functions are now tightly integrated. The boom of leisure activity has given tourists higher expectations than ever. They expect airports to be destinations in themselves, with multiple facilities where they can enjoy memorable experiences. China is a country with such an enormous amount of culture, cuisine, literature and climate, and there is a huge difference between its cities, such as Xi'an, Changsha or Guiyang. So each city has a different challenge when it comes to the airport. China's 13th Five-Year Plan focused on China's aviation industry, grabbing attention from the entire world. This current vision encourages Aedas to continuously move forward.

6. China's Megacity and Urban Design

Urban development is almost complete in Beijing, Shanghai and Guangzhou. What does Aedas think of the way they have developed?

China already has the world's three largest megapolises — the Pearl River Delta (population of 120 million), Yangtze River Delta (88 million) and the Bohai Economic Rim (66 million).
We started to see China recovering from its economic slowdown, especially in inland cities, and we have accepted commissions for many urban re-planning projects, which are quite different from what we did for new cities in 2012 and 2013. These urban re-planning, or perhaps better called urban regeneration, is the major trend in developed cities. Of course, this trend is interconnected with other things like city hubs. China has already embarked on creating the world's highest density, high-rise smart cities which will address the needs of our modern internet-connected live-work society. Take Shanghai for instance. Today, the residential and business areas are mostly separated. By building city hubs upon existing infrastructure nodes, it will develop high density live-work communities that have high value land with good connectivity, densification and better use of older development among existing cities.

What does Aedas expect from developing urban agglomerations such as Hong Kong-Shenzhen-Guangzhou, Chongqing-Chengdu, Shijiazhuang-Zhengzhou and so forth? What is Aedas' strategy for dealing with them?

Aedas is a local and international company that does not view itself as a foreign company in any city for which it designs. In order to understand local culture, climate and economy, we establish offices in each major region for which we design. Our offices in Beijing, Shanghai, Chengdu, Shenzhen and Hong Kong understand the unique nature of north, east, west and south China, respectively, and deliver this knowledge to our designers in whichever of the five Chinese cities they are located. Most importantly, all of Aedas' China design is carried out within these five offices due to the need to understand the particular needs of China and the wealth of high-rise, high density design experience held in these offices. We will work on commissions in the developing regions from these offices, and if needed, expand our office network to more Chinese cities.

7. Culture and Story

Local identity is one of the highlights in Aedas' projects. How does Aedas embed local culture into design?

Aedas has a great understanding of China through the 800 staff in our five China offices. In our projects, Aedas seeks to understand local culture, geography, history and economics and to embed this understanding into our designs so that they are relevant and powerful. We are particularly concerned about embedding local culture, history and legend into our designs. This makes each of them unique to its city, location and culture. More importantly, we use the building designs to recall culture, past events and geography so that they have great relevance to the city, bringing to mind wonderful and interesting stories for generations of people to appreciate.

Aedas respects the continuity of historical context. On the site of Center 66 in Wuxi, we found a two-story opera house flanked by ancillary stages in a pavilion and a chamber dating from the Ming Dynasty in the 14th century. The design preserved these stages. The requirement of preserving these buildings basically drove the whole master plan of the project. We tried to group these buildings, along with Qian Zhongshu's house and an old well, using a central axis. We integrated the big public landscape square with the preserved buildings, creating an organic architectural complex. This boosts the communication between new civic space and historic architecture.

What new elements has Aedas embedded in its designs in order to narrate a story?

The history embedded in the historical buildings or the architectural form is a very good starting point for storytelling. As far as newly built architecture is concerned, we need to find some references and use metaphor or symbolism to articulate. Clients can easily understand function. Form is more abstract and subjective, which is about aesthetics — about how to integrate shape, colour and materiality. Therefore, it is necessary to employ metaphor to help people understand and appreciate the architectural form.

In the case of Olympia 66 in Dalian, we designed the exterior form based on functionality and endowed the design with the concept of "twin carp." This immediately gained acknowledgment and appreciation from local government and client. From 2008 to 2009, architecture was popularised in the public eye with buildings such as the so-called "Water Cube" and "Bird's Nest". When architecture contains such concepts, such good messaging, and such symbolism, it can have a deeper impact on civic consciousness. These are the elements that give architecture its legacy.

8. Belt and Road

Please talk about the Belt and Road, especially the Maritime Silk Road. And please elaborate on how Aedas will respond to the Belt and Road in urban planning and the way it undertakes its projects?

The Maritime Silk Road spans from South China through Southeast Asia and Africa to Europe over the sea. Chinese developers' projects such as Greenland's in Iskandar, and China Merchants' in the ports of Djibouti, are all along the trade route. In Hong Kong, The Forum demonstrates how a park can be elevated above ground and connected to other programs in the heart of a central business district, setting an example of multi-level public spaces; Hong Kong International Airport Midfield Concourse and North Satellite Concourse offer modern day travellers' experience with high flexibility to adapt to the future. In Singapore, The Star breaks the rule of building air-conditioned malls in tropical countries, and both The Star and Sandcrawler extend the public realm. We have just completed Unilever Headquarters in Indonesia. Its design successfully embodies the client's vision, values and blends their European tradition with the Indonesian context. In Dubai, the Dubai Metro is the first metro system in the Gulf and the benchmark for future rail development in the region. The Djibouti Harbour Regeneration Master Plan regenerates the historic port of Djibouti by relocating and expanding the port facilities to attract foreign investment to the city and wider country. In Iskandar, principles of high-density live-work are applied in a new Malaysian town based on our China experience. In London, the regeneration of the Regent Quarter has brought dynamism and an economic boost to the area.

Will Aedas share the experience of China with the world?

Chinese cities possess high-density environments with an enormous group of people living in high-rise buildings. We have gained unique design experience from that. China initially implemented American notions of what developments should be, later extending to a Hong Kong-style commercial multi-complex development pattern. But in 2008, China developed its own unique design that works in numerous cities, cultures and climate, carrying forward a new way of designing buildings and cities.

Quite a lot of Western architects do not understand high-density or high-rise buildings. University students in architecture in New York and London could not figure out how many elevators are needed in a building. Also, they could not draw a semblance of the structure. This is because building a high-rise building is very difficult in New York and London. The towers in London are very expensive and very inefficient and take long time to build. The value of those towers lies more in their sculptural quality. London could not be regarded as high-density city since its 220 high-rise buildings are spread out all over the city. The poor infrastructure network is not capable of sustaining mass transportation between the separated hubs. Therefore, what London needs to consider is how to distribute these towers and the density. But many cities in China are more efficient and energetic since density is concentrated. This reminds me of the high-density development of the Shanghai metro system, which has provided the public a more convenient life. Through research into the subway system, the development of Shanghai's city hubs has allowed us to predict the land's value. Where two to three subway stations come together, the land is supposed to be of great value and will be developed into a future city hub. This is a good example of Chinese experience teaching the world. Just as in the manufacturing industry, the experience of China can be shared with the world. For instance, in a tall office building, the efficiency of workspaces is effected by the size of the core. So, we have to make the core as small as we can. Some elevator manufacturers are trying to find ways to make elevators more efficient. They try to put two elevator cars into the same shaft, even integrating smart phone control. Numerous systems are being developed with the Chinese market in mind. These technologies and experience are worth sharing with other cities around the world.

9. Urban Renewal and Restoration

What does Aedas think about adaptive reuse project?

Aedas' deep understanding of local situations is well reflected in our project research. It has also provided suggestions to local governments and landowners on how they can renew and improve their properties. We would like to take Hong Kong's central business district as an example. Aedas has been working in Central for almost 30 years, with projects including Lan Kwai Fong and The Landmark, a commercial and office center. Aedas revitalised, refurbished and rebuilt some major buildings in Central, which is known as Asia's Wall Street. We transformed Central to an all-day and multi-layered live-work-recreate city center. We are well aware of market needs, local culture and changes. We transformed a place that only served banking duties into one of the world`s most successful central business districts with retail, hotels, offices, leisure and entertainment. As architects, we are concerned about urban environments in exactly these ways.

We would also like to take historic theatres as examples. When we began spearheading the refurbishment and regeneration of historic British theatre buildings in the 20th century, attitudes towards conservation and conservation practice were very different indeed. At that time Victorian and Edwardian theatres (built around 1860 to 1915) were poorly regarded by the general public, planning authorities and the architectural community. This lack of interest in theatres at that date was all pervasive and extended to many other typologies, with notable buildings and elements of town planning lost as a result. In this context, anyone looking to salvage an historic theatre needed to seek a lot of creative licenses approved by the local and national planning regulations. This can be seen in our work on Nottingham's Royal Concert Hall. The original theatre dates from 1865 and refurbishment was undertaken in 1978. To the untutored eye, the theatre looks like a well preserved theatre, but the architectural treatment made significant interventions and alterations. There was a clear distinction made between the treatments of historical performance space and public areas, where a more contemporary design language was used. There was a sensitivity to the historic value of the performance space, but beyond this theatrical environment, modern audiences were given contemporary public spaces. Moving forward 40 years, with the increased awareness in the value of our historic buildings and changing public attitudes, the way we treat and retrofit historic theatres has changed markedly. Fewer historic theatres remain, and some remain at risk — so their inherent value as a typology increases. Today we work within tighter regulations and we are required to consult closely with a number of local and national historic and conservation institution.

A tool for all decision making is the detailed conservation statements that are prepared for each building. They analyse the historical value and development of a theatre, looking at additions and subtractions, and at its social importance and cultural significance. As a result, we, as designers, make much more informed decisions regarding the redevelopment strategy. We understand much more about historic materials and construction techniques and can make educated value judgments when we balance the historic fabric against the requirements of modern day entertainment and audience expectations. Our design and rejuvenation of historic theatres has undoubtedly been enhanced over the last 40 years — as with other designers working in other fields of conservation.

What stance does Aedas take in building restoration, especially in specific historic buildings?

It depends. Taking Center 66 in Wuxi as example, we preserved the Chenghuang Temple Precinct and relocated it to the heart of a large public plaza within the development. The cluster comprises a double-storey opera house flanked by ancillary stages in a pavilion and a chamber dating from the Ming Dynasty in the 14th century. The former opera house had been badly damaged over the years. The compound was faithfully restored to its original splendor, including structural underpinning, foundation strengthening and roof repairing. The open landscaped plaza wrapping around the compound emphasises the buildings' historical resonance while creating an emphatic contrast with their contemporary setting.

The Art Community: Revitalisation Project at Mallory Street and Burrow Street in Hong Kong, on the other hand, revitalises a rare and intact ensemble of historic shophouses and turns them into a modern venue for public leisure and the cultural and creative industries. Some of the elements are restored with modern-day technologies, while some are completely new parts. By deploying innovative design, we reinterpreted

Cast 为英国唐卡斯特的新建演艺场所，是 MUSE Waterdale（位于唐卡斯特中心的市民文化区）总体规划的核心元素
Cast is Doncaster's new performance venue and is a key element of MUSE Waterdale Masterplan proposal for a civic and cultural quarter in the heart of Doncaster, UK

traditional spatial wisdom and use of materials, and upgraded the original structure and facilities to comply with modern building regulations and needs so that the architectural mission of the shophouses can be sustained.

In the Regent Quarter, we engaged in adaptive reuse and restoration, as well as revitalisation by inserting new buildings. In the British arts sector, restoration and refurbishment are considered and costed before the option of a new build is sought. If a building is in need of refurbishment, redevelopment or replacement, then a wide range of public consultations takes place to ensure the project will increase public usage and maximise a range of social values, not just land values. The bottom line of all our restoration and new-build projects is called "the quadruple bottom line": maximising social, cultural, economic and sustainability values across the life of the project and beyond.

Taking Cast in Doncaster as an example. Different theatres and clients demand different approaches; buildings can be restored timelessly, or become a sequence of spaces characterised by seamless alterations and additions, or become more avant-garde in their imagination. What unites these different approaches in our work is that they all share an appreciation of how historic building fabric can inspire the creative process of making theatre — and the civic pride these valuable pieces of our collective cultural heritage bring to our communities.

China, like the UK, has a number of active organisations that dedicate themselves to protecting historic buildings, and buildings of significance. In the UK these organisations range from the national Historic England and Theatres Trust groups, to the local government level with local conservation officers. They also have a number of other focus groups such as the 20th Century Society, Victorian Society and so on, all dedicated to preserving history where they can. When restoring a historic building in the UK, permission needs to be granted from these organisations before work can commence.

The UK has been protecting 20th century buildings for some time now, however China has only recently done this with the publishing of the first 20th Century Architectural Heritage List in 2016. There is a great deal we can learn from each other, in regards to restoring historical buildings, especially in the area of expert crafts skills. With the proliferation of prefabricated buildings, the skill base of craftspeople such as stonemasons, thatchers, wood carvers, traditional plasterers, is in decline. Restoring these buildings can help save these skills.

Aedas 在中国精选项目分布图
Aedas Selected Projects in China

D 一带一路 BELT AND ROAD

NO.29 新加坡星宇项目 P210
The Star, Singapore

NO.30 新加坡 Sandcrawler P216
Sandcrawler, Singapore

NO.31 阿联酋迪拜 Ocean Heights P222
Ocean Heights, Dubai, UAE

NO.32 阿联酋迪拜地铁站 P226
Dubai Metro, Dubai, UAE

NO.33 英国唐卡斯特 Cast 剧院 P232
Cast, Doncaster, UK

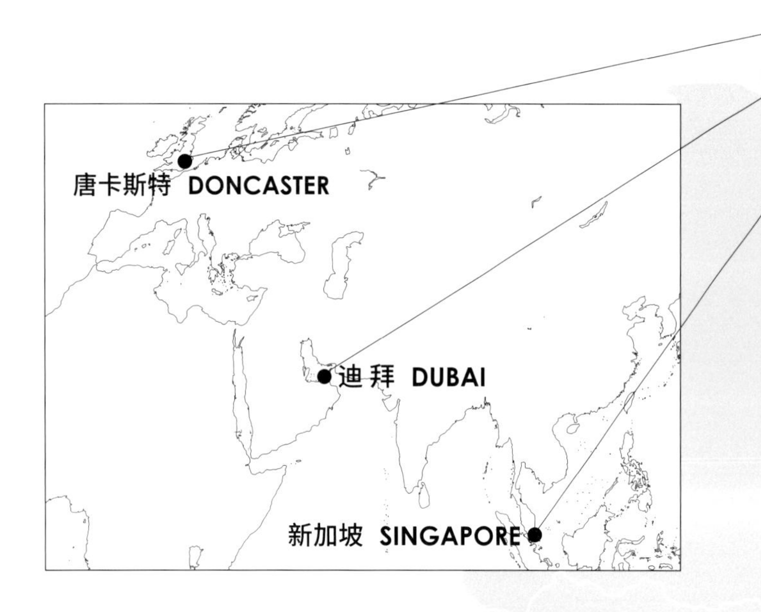

B 中国大陆长江以南 SOUTH OF YANGTZE RIVER

NO.10 上海星荟中心 P096
Shanghai Landmark Center, Shanghai

NO.11 上海龙湖虹桥项目 P102
Longfor Hongqiao Mixed-use Project, Shanghai

NO.12 上海虹桥世界中心 P108
Hongqiao World Center, Shanghai

NO.13 无锡恒隆广场 P112
Center 66, Wuxi

★ NO.14 苏州西交利物浦大学中心楼 P118
Xi'an Jiaotong-Liverpool University Central Building, Suzhou

NO.15 义乌之心 P126
The Heart of Yiwu, Yiwu

★ NO.16 广州南丰商业、酒店及展览综合大楼 P130
Nanfung Commercial, Hospitality and Exhibition Complex, Guangzhou

NO.17 广州邦华环球贸易中心 P136
Bravo PARK PLACE, Guangzhou

NO.18 珠海粤澳合作中医药科技产业园总部大楼 P142
Headquarters, Traditional Chinese Medicine Science and Technology Industrial Park of Co-operation between Guangdong and Macao, Zhuhai

NO.19 深圳宝安国际机场卫星厅 P146
Shenzhen Airport Satellite Concourse, Shenzhen

★ NO.20 珠海横琴国际金融中心 P150
Hengqin International Financial Center, Zhuhai

NO.21 珠海横琴中冶总部大厦（二期） P158
Hengqin MCC Headquarters Complex (Phase II), Zhuhai

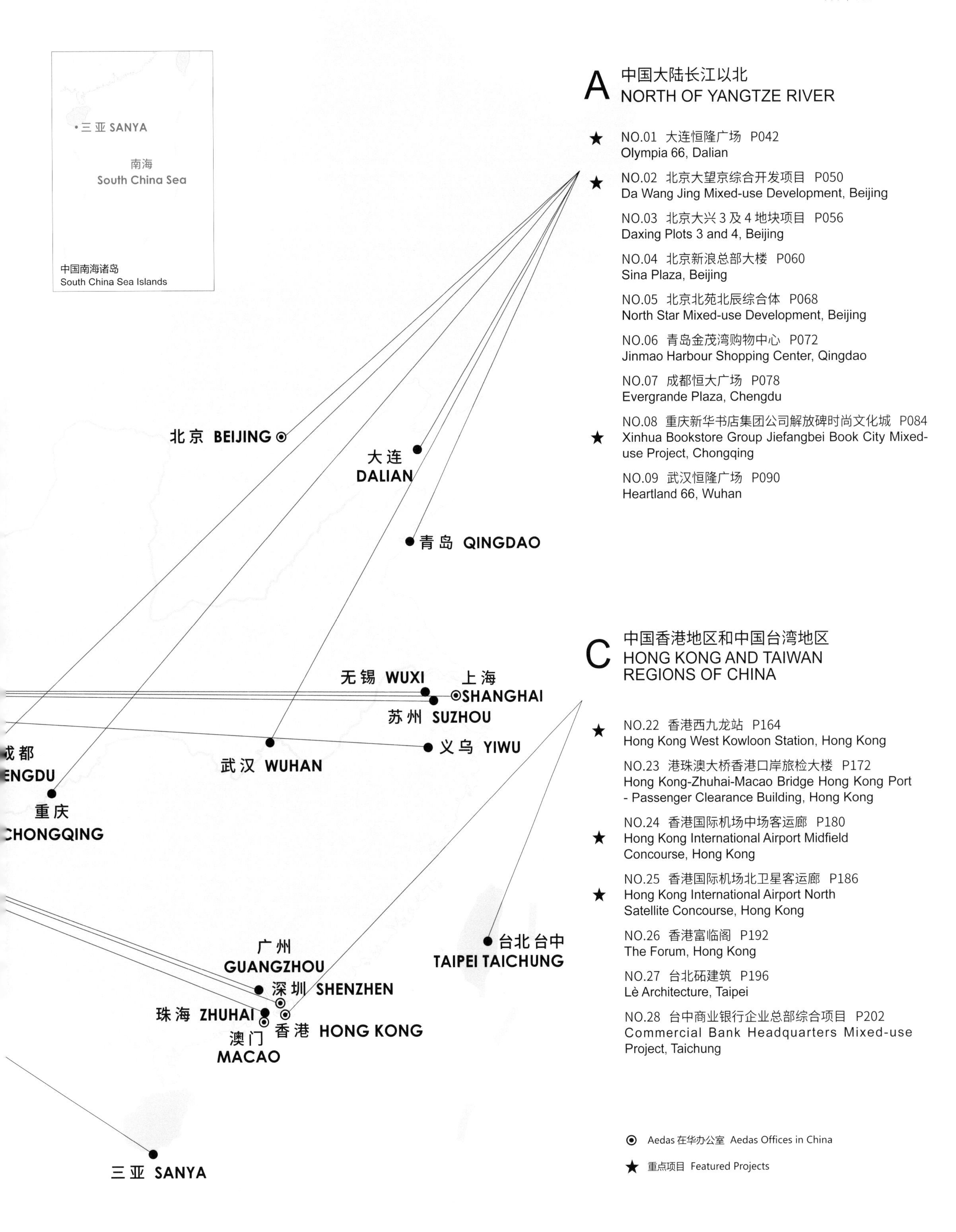

A 中国大陆长江以北 NORTH OF YANGTZE RIVER

★ NO.01 大连恒隆广场 P042
Olympia 66, Dalian

★ NO.02 北京大望京综合开发项目 P050
Da Wang Jing Mixed-use Development, Beijing

NO.03 北京大兴 3 及 4 地块项目 P056
Daxing Plots 3 and 4, Beijing

NO.04 北京新浪总部大楼 P060
Sina Plaza, Beijing

NO.05 北京北苑北辰综合体 P068
North Star Mixed-use Development, Beijing

NO.06 青岛金茂湾购物中心 P072
Jinmao Harbour Shopping Center, Qingdao

NO.07 成都恒大广场 P078
Evergrande Plaza, Chengdu

★ NO.08 重庆新华书店集团公司解放碑时尚文化城 P084
Xinhua Bookstore Group Jiefangbei Book City Mixed-use Project, Chongqing

NO.09 武汉恒隆广场 P090
Heartland 66, Wuhan

C 中国香港地区和中国台湾地区 HONG KONG AND TAIWAN REGIONS OF CHINA

★ NO.22 香港西九龙站 P164
Hong Kong West Kowloon Station, Hong Kong

NO.23 港珠澳大桥香港口岸旅检大楼 P172
Hong Kong-Zhuhai-Macao Bridge Hong Kong Port - Passenger Clearance Building, Hong Kong

★ NO.24 香港国际机场中场客运廊 P180
Hong Kong International Airport Midfield Concourse, Hong Kong

★ NO.25 香港国际机场北卫星客运廊 P186
Hong Kong International Airport North Satellite Concourse, Hong Kong

NO.26 香港富临阁 P192
The Forum, Hong Kong

NO.27 台北砳建筑 P196
Lè Architecture, Taipei

NO.28 台中商业银行企业总部综合项目 P202
Commercial Bank Headquarters Mixed-use Project, Taichung

◎ Aedas 在华办公室 Aedas Offices in China

★ 重点项目 Featured Projects

中国大陆长江以北
NORTH OF YANGTZE RIVER

中国大陆长江以北 NORTH OF YANGTZE RIVER

- ❷❸❹❼ 大连恒隆广场
  Olympia 66, Dalian
- ❷❹❻ 北京大望京综合开发项目
  Da Wang Jing Mixed-use Development, Beijing
- ❸❹ 北京大兴 3 及 4 地块项目
  Daxing Plots 3 and 4, Beijing
- ❶❸❻ 北京新浪总部大楼
  Sina Plaza, Beijing
- ❶❷❸❹❻ 北京北苑北辰综合体
  North Star Mixed-use Development, Beijing
- ❷❸❹❼ 青岛金茂湾购物中心
  Jinmao Harbour Shopping Center, Qingdao
- ❶❷❹ 成都恒大广场
  Evergrande Plaza, Chengdu
- ❶❼ 重庆新华书店集团公司解放碑时尚文化城
  Xinhua Bookstore Group Jiefangbei Book City Mixed-use Project, Chongqing
- ❷❸❹❼ 武汉恒隆广场
  Heartland 66, Wuhan

中国大陆长江以南 SOUTH OF YANGTZE RIVER

- ❸❻❼ 上海星荟中心
  Shanghai Landmark Center, Shanghai
- ❶❷❸❹❻ 上海龙湖虹桥项目
  Longfor Hongqiao Mixed-use Project, Shanghai
- ❻❼ 上海虹桥世界中心
  Hongqiao World Center, Shanghai
- ❷❸❹❻❼ 无锡恒隆广场
  Center 66, Wuxi
- ❸❼ 苏州西交利物浦大学中心楼
  Xi'an Jiaotong-Liverpool University Central Building, Suzhou
- ❷❸ 义乌之心
  The Heart of Yiwu, Yiwu
- ❷❸❻ 广州南丰商业、酒店及展览综合大楼
  Nanfung Commercial, Hospitality and Exhibition Complex, Guangzhou
- ❸❽ 广州邦华环球贸易中心
  Bravo PARK PLACE, Guangzhou
- ❸ 珠海粤澳合作中医药科技产业园总部大楼
  Headquarters, Traditional Chinese Medicine Science and Technology Industrial Park of Co-operation between Guangdong and Macao, Zhuhai
- ❸❽ 深圳宝安国际机场卫星厅
  Shenzhen Airport Satellite Concourse, Shenzhen
- ❶❼ 珠海横琴国际金融中心
  Hengqin International Financial Center, Zhuhai
- ❷❸❹ 珠海横琴中冶总部大厦（二期）
  Hengqin MCC Headquarters Complex (Phase II), Zhuhai

中国经历改革开放四十年的发展，城市建设基本完成，现在站在了新的起点，社会经济要转变增长方式。北方从原来的工业化城市转变为新型现代化城市，有新经济互联网带来的变化，也有国家战略促成的机遇，京津冀联动，功能产业重新布局对城市建设影响深远。Aedas 设计的北京新浪总部大楼、大连恒隆广场等开启了城市建设发展的创新大门。

After 40 years of reform and opening up, China has completed most urban construction and is now standing at a point to find new ways to grow the economy. The transformation of northern China from industrialised cities to modern cities is the result of the rising of internet industry and promoting by national policies. The Beijing-Tianjin-Hebei collaboration and the rising of new industries have long-term implications to urban development. Aedas-designed projects such as Beijing Sina Plaza and Dalian Olympia 66 have opened the door to innovative urban development.

中国香港地区和中国台湾地区 HONG KONG AND TAIWAN REGIONS OF CHINA

- 1 5 8 香港西九龙站 Hong Kong West Kowloon Station, Hong Kong
- 2 3 5 6 港珠澳大桥香港口岸旅检大楼 Hong Kong-Zhuhai-Macao Bridge Hong Kong Port - Passenger Clearance Building, Hong Kong
- 1 5 8 香港国际机场中场客运廊 Hong Kong International Airport Midfield Concourse, Hong Kong
- 1 5 8 香港国际机场北卫星客运廊 Hong Kong International Airport North Satellite Concourse, Hong Kong
- 4 香港富临阁 The Forum, Hong Kong
- 3 7 台北砳建筑 Lè Architecture, Taipei
- 3 台中商业银行企业总部综合项目 Commercial Bank Headquarters Mixed-use Project, Taichung

一带一路 BELT AND ROAD

- 3 4 8 新加坡星宇项目 The Star, Singapore
- 3 8 新加坡 Sandcrawler Sandcrawler, Singapore
- 8 阿联酋迪拜 Ocean Heights Ocean Heights, Dubai, UAE
- 5 8 阿联酋迪拜地铁站 Dubai Metro, Dubai, UAE
- 8 9 英国唐卡斯特 Cast 剧院 Cast, Doncaster, UK

九项设计理念 NINE POINTS OF DESIGN IDEAS

1. 《国家新型城镇化规划（2014—2020 年）》 NATIONAL NEW URBANISATION PLAN (2014-2020)
2. 高密度城市枢纽 HIGH DENSITY CITY HUBS
3. 具有通透性和连接性的设计 POROUS AND CONNECTED DESIGNS
4. 新型商业零售 THE NEW RETAIL
5. 基础设施设计 INFRASTRUCTURE DESIGN
6. 中国的特大城市与城市设计 CHINA MEGAPOLIS AND URBAN DESIGN
7. 文化和故事 CULTURE AND STORY
8. 一带一路 BELT AND ROAD
9. 城市改造与修复 URBAN RENEWAL AND RESTORATION

大连恒隆广场

Olympia 66, Dalian

场地策略：以退为进，融入周边

每一个场地背后，都有其特定的文化背景。每一个设计背后，都不应脱离地域来解构。大连，从小渔村发展为东北地区最重要的城市之一，如今在市区建立起一座超越购物中心功能的标志性建筑——大连恒隆广场，Aedas 将其定义为“一次地域文化的深耕实践”。除了被定位为大连市最大的购物中心，这座建筑还身兼成为代表城市气质的地标的责任，这是业主与当地政府共同的诉求。

大连恒隆广场坐落于大连商业及住宅区中心，原址为人民体育场，周边有一系列广场穿过项目所在的中心城区，其中与项目相邻的奥林匹克广场一直延伸至大连市政府办公大楼前方的亚洲最大的广场——星海广场。在现有的城市规划格局中，建筑被限高在 60m 之内。对于这样一个处于城市腹地重要位置的限高用地，Aedas 除了考虑建筑与周边环境的整体性之外，在最后的设计方案中以退为进，将建筑平铺于整个基地，外观采用多个建筑元素立体拼接，与周边的建筑群对话，亦庄亦谐。

1. 大连城市景观
Dalian city
2. 北侧建筑外观
External view of the north elevation
3. 建筑外观
Aerial view

项目信息 | PROJECT INFORMATION

地　　点：中国大连
建筑面积：222 000m²
建成时间：2015 年
项目功能：商业零售
业　　主：恒隆地产有限公司
主要设计人：林静衡，祈礼庭

Location: Dalian, PRC
Gross floor area: 222,000 m²
Completion year: 2015
Sector: Retail
Client: Hang Lung Properties Ltd.
Directors: Christine LAM, David CLAYTON

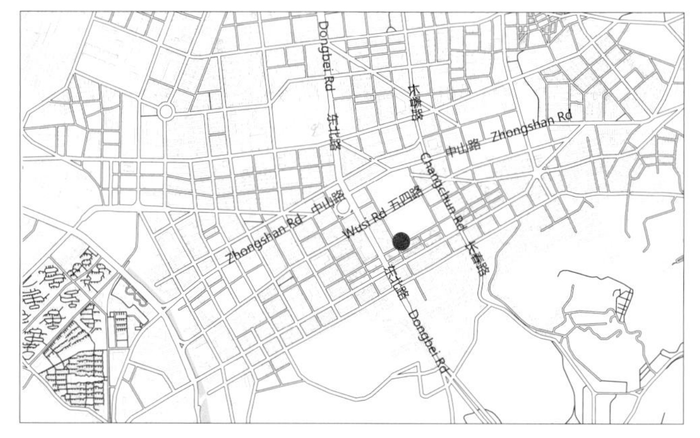

设计理念：追本溯源，回应中国文化

地标建筑应该是地域文化的积淀与延续。从小渔村到国际知名港口，“渔”是大连的“关键词”之一。正如 Aedas 一贯将国际化与本土文脉结合的设计策略，大连恒隆广场的建筑设计灵感源自于“渔”，同时又借鉴了中国农历新年剪纸艺术中双鱼追尾嬉戏的造型：代表着鲤鱼轮廓的多个相交弧深化成一组图案，它们相互融合，代表着流动、平衡和繁荣兴旺。方案用两个相连的椭圆勾勒出鲤鱼主题，两个椭圆偏移 45°，沿相邻的边缘延伸，在两个椭圆上放置第三个椭圆，以加强两者之间的联系。高空俯瞰，大连恒隆广场像两条灵巧的鲤鱼首尾相随，围绕商场中央游动，巧妙地传达出“太极双鲤”的吉祥之意。从双鲤在水中畅游扭动的曼妙姿态出发，Aedas 塑造了一个优美独特、又能回应城市文脉的几何建筑形体。

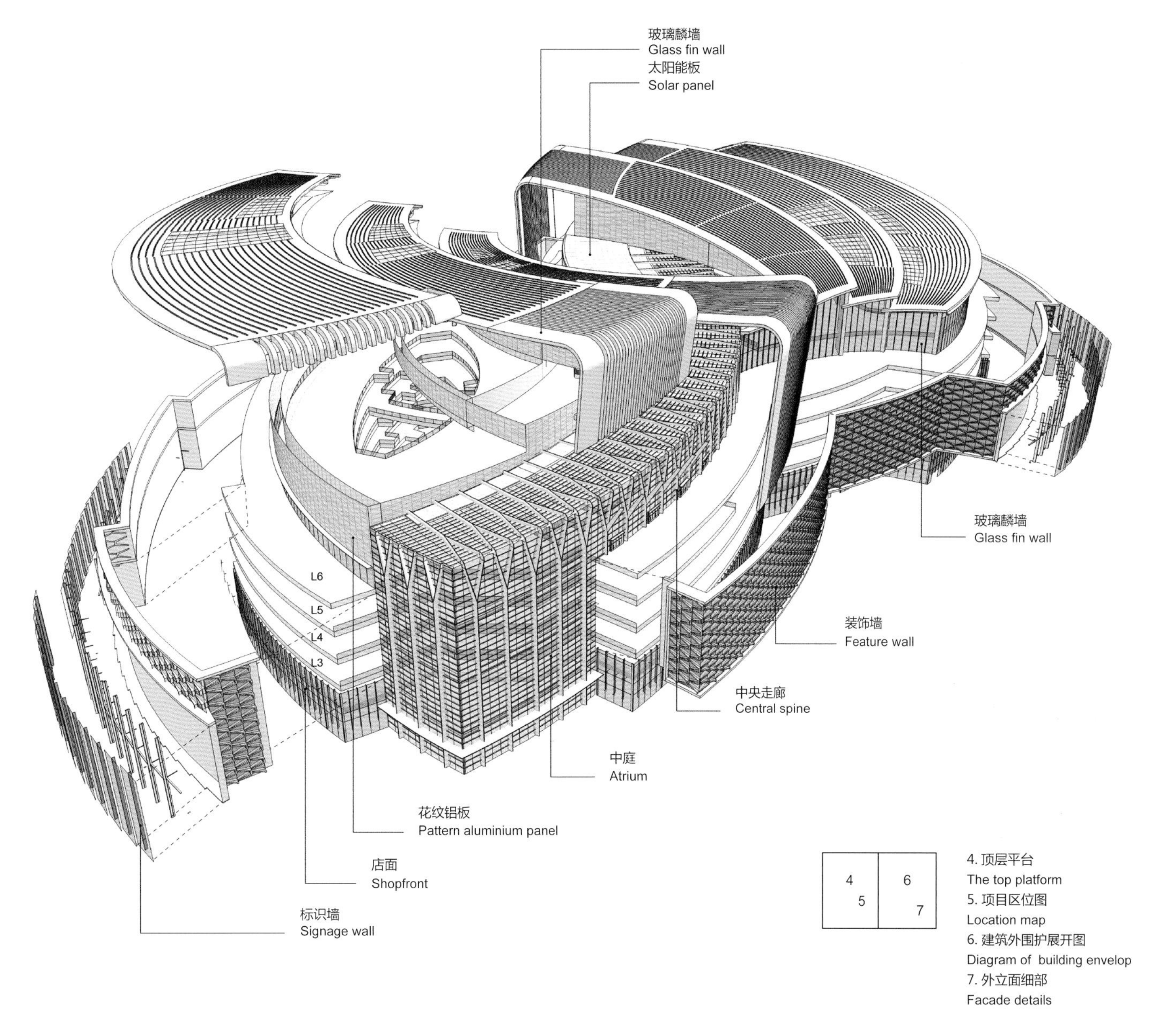

4 5	6 7

4. 顶层平台
The top platform
5. 项目区位图
Location map
6. 建筑外围护展开图
Diagram of building envelop
7. 外立面细部
Facade details

功能塑造：打造全天候休闲目的地

商场应是令人流连忘返的全天候休闲目的地。

为了塑造大连恒隆广场购物中心，Aedas 用简明的几何平面布局为顾客提供清晰的方向感，同时，建筑通过设置多个入口加强了与城市的连系。街面层连续的临街铺面不仅极大地提升了商业零售店的可见性，还赋予首层极佳的渗透性。主入口位置突出，且开拓出宽敞的公共活动空间，令来访者在进入购物动线之前可驻足停留。宽敞的楼面被划分为开放的公共活动空间和商业零售区域，中央走廊连接两侧巨大的中庭，并以对角线的方式连接位于角落的入口。各个休闲活动和公共活动空间沿商业零售路线排列，让人们随时能够在咖啡馆休憩或观看展览。围绕建筑的绿化景观平台更营造出轻松的购物氛围。

在 Aedas 的操刀下，大连恒隆广场完全满足了业主与当地政府的诉求——超越购物中心功能的地标建筑。

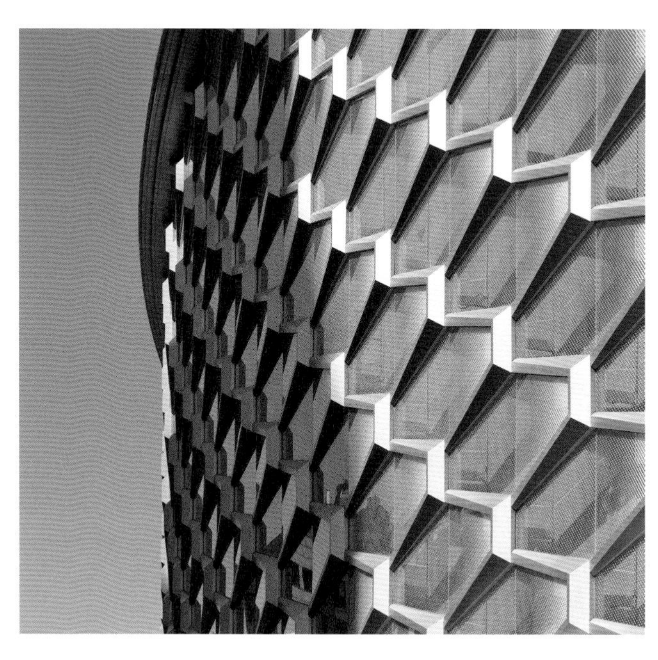

Master Plan Strategy: Retreat to Move Forward and Merge with the Surroundings

Every site has its unique cultural background. No design should be created without consideration to the cultural characteristics of the region in which it finds itself. Dalian, which has grown from a small fishing village to one of the major cities in northeastern China, has constructed Olympia 66, a new iconic landmark which is more than just a shopping mall. Aedas defines Olympia 66 as "a practice based on a deep dig into regional culture." Not only is Olympia 66 the largest shopping mall in Dalian, it has the responsibility of representing the city's character. This was a requirement imposed by both the clients and the local government.

Olympia 66 is located on the site of the former People's Stadium, with a sequence of squares running through this central zone. Adjacent to the project is the Olympic Square which extends towards Xinghai Square in front of the local government center, which is reputedly the largest civic square in Asia. The Dalian government imposes a 60 metres height restriction on this site. The response to this challenge was to retreat to move forward — to build over the whole of the site and treat its external design as a collection of three-dimensional architectural elements montaged together. Some are striking, while others more neutral, but they are always dramatically connected to the neighbouring buildings.

Design Concept: Trace and Respond to Chinese Culture

Iconic architecture needs to find its inspiration and relevance in local culture. Fish is one of the keywords that defines Dalian, as a well-known international port that was once a small fishing village. Aedas combines global perspectives and local context, and draws inspiration from the "twin carps" symbol typically used in Chinese New Year paintings or paper-cuts. The pattern consists of interweaved curving arcs which resembles the contours of the carp, symbolising dynamism, balance and wealth. The design uses

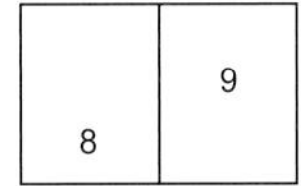

8. 中央走廊
Central spine
9. 内部中庭
Atrium

维苏威
PIZZA
TIGERTOPOKI

two superimposed ovals to outline the carp theme, then applies a third one to strengthen the relationship between the previous two ovals. When it is seen from above, Olympia 66 looks exactly like twin carps chasing each other around the center of the mall, conveying the meaning of luck in a subtle way. Drawing inspiration from the graceful posture of the swimming fishes, Aedas has created a beautiful and unique architectural form which fits harmoniously into the urban environment.

Programme: Creating an All-day Leisure Destination

With regards to the organisation of the shopping center, Aedas has provided a clear layout that can direct visitors to destinations within the building. The building form enhances urban connectivity and integration through multiple entrances. There is a continuous shop frontage at the street level, which not only provides great visibility for retail tenants, but also makes the building totally permeable at ground level. Main entrances are prominently defined and open into generous event spaces, allowing visitors to pause before joining the vortex of circulation. The enormous floor area is divided into open event spaces and retail areas with a central spine that runs through the center of the mall, linking two large atria on each side and diagonally linking the corner entrances. Leisure activities and event spaces are sequenced along the retail route, providing opportunities for visitors to rest at a cafe or take part in an exhibition. Landscape terraces around the building further enhance a relaxing shopping environment.

Aedas has created exactly what the client and local government wanted: an iconic facility that transcends its overall function as a shopping mall.

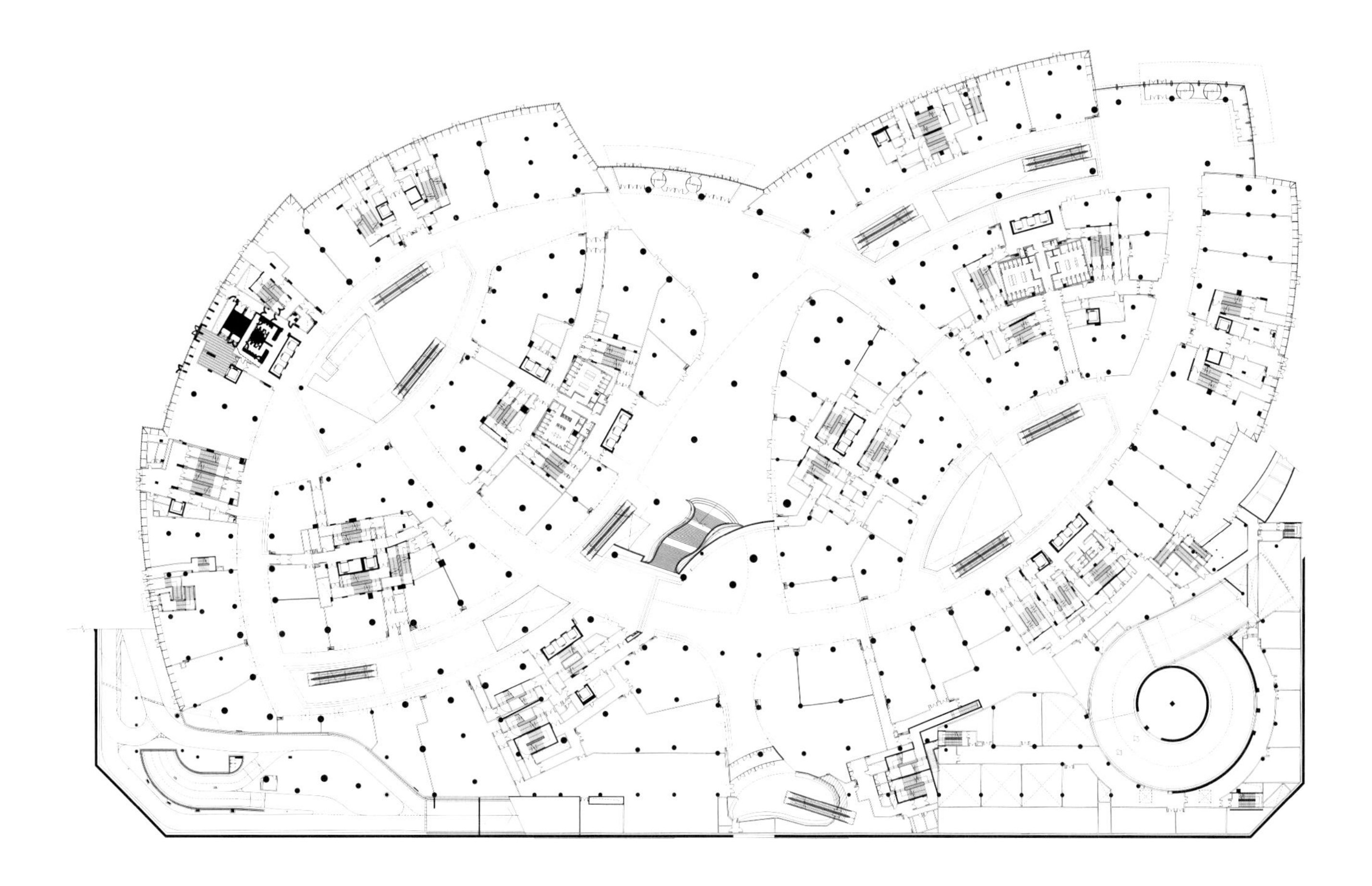

10	11
	12
	13

10. 滑冰场
Skating rink
11. 一层平面图
Level 1 floor plan
12. 南立面图
South elevation
13. 剖面图
Section

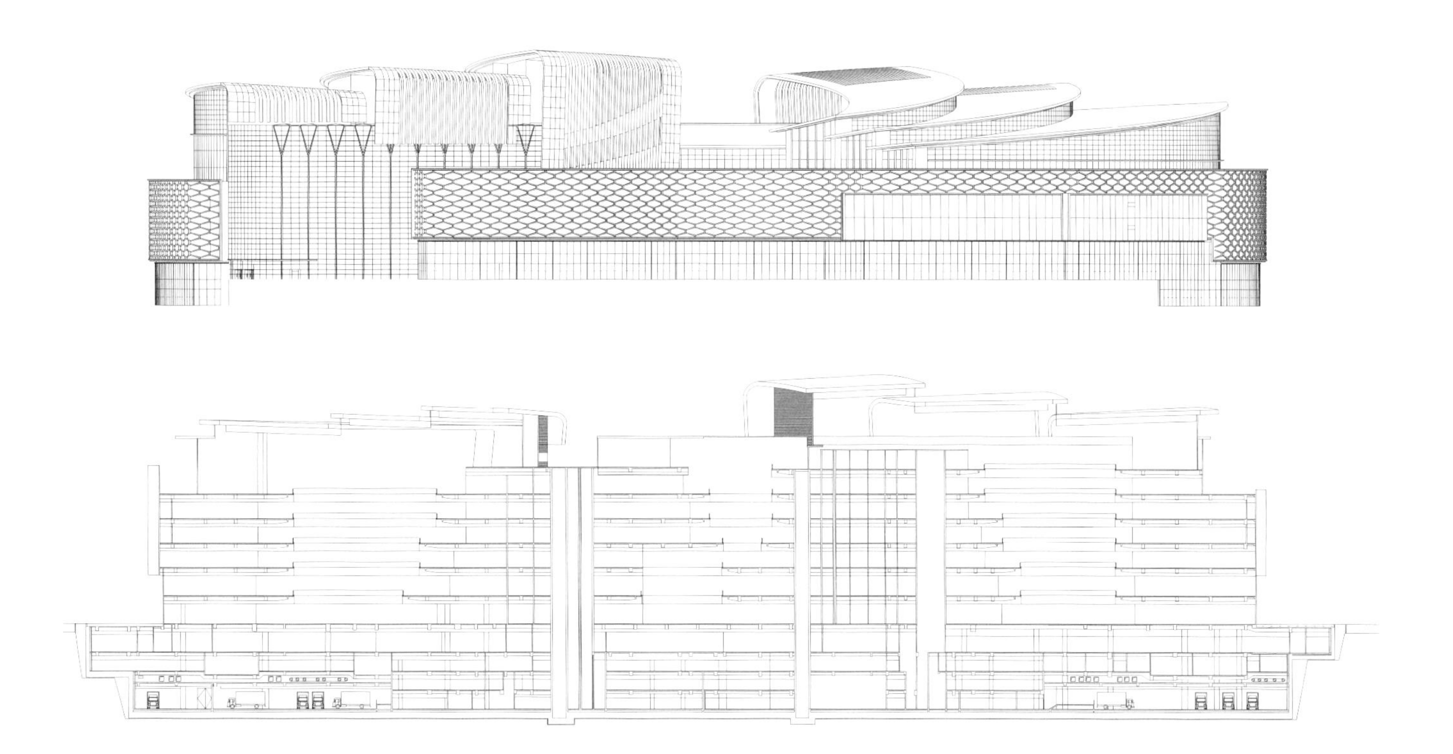

北京大望京综合开发项目

Da Wang Jing Mixed-use Development, Beijing

大望京综合开发项目地处北京东北区域的门户地区，包括三个独立的地块，总体规划面积近 600 000m²。基地北面是地铁 15 号线的望京东站，东北角毗邻大望京公园和东五环路。作为“大望京商务区”的重要组成部分，项目共由 4 栋超 5A 高层写字楼、1 栋高层商务公寓和 1 座多功能商业会展大楼组成。超高层建筑的裙楼中也设置了各类附属设施，包括私人及企业的商业会所、餐饮、零售、银行和娱乐等空间。

基地周围聚集着住宅和超高层建筑，用地情况复杂，因此，设计在对周围城市风貌和建筑体量进行建模、模拟的基础上，进行了多轮方案对比，最终得出理性而高效的总平面，最大程度降低了周边建筑对本项目视野的遮挡。项目以“绿洲”为核心概念，规划强化了基地与周边绿化环境的关系，将多个地块紧密整合在一起，连接起周边的公园空间，达到了 40% 绿化率。

建筑设计上，5 栋 160~220m 的超高层主楼和附属裙楼的外立面和表皮，均突显了有机与现代相结合的建筑理念。而考虑到建筑本身复杂、独特的造型，以及空间高使用率的严格要求，设计从概念、方案到施工阶段，都采用了 Revit 与 Rhino 相结合的参数化设计手段。

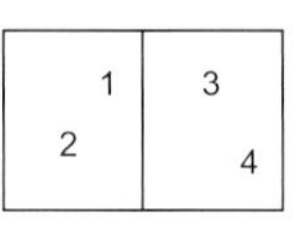

1. 大望京商务区景观
Cityscape of Da Wang Jing business district
2. 建筑外观
External view
3. 大望京综合开发项目全景
Overview of the project
4. 项目区位图
Location map

项目信息 | PROJECT INFORMATION

地　　点：中国北京
建筑面积：571 878m²
设计时间：2012—2014 年
建成时间：2018 年
项目功能：办公楼
业　　主：北京乾景房地产开发有限公司
主要设计人：Andrew BROMBERG

Location: Beijing, PRC
Gross floor area: 571,878 m²
Design time: 2012-2014
Completion year: 2018
Sector: Office
Client: Beijing Qian Jing Real Estate Development Ltd.
Design Director: Andrew BROMBERG

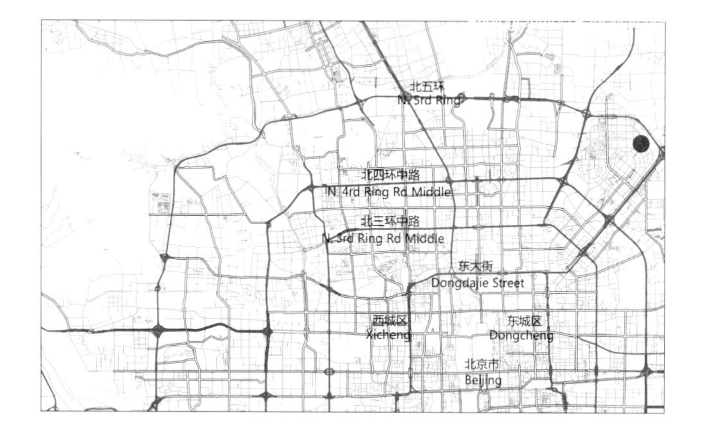

Da Wang Jing Mixed-use Development in northeast Beijing consists of three land parcels with a planned area of more than 600,000 square metres. The site is bounded by Wang Jing East subway station in the north and Da Wang Jing Park and the East Fifth Ring Road in the northeast. As an important component of the Da Wang Jing business district, the project includes four Grade-A supertall office towers, a serviced apartment tower and a multi-functional commercial exhibition complex. The podiums of the high-rises also offer a variety of auxiliary facilities, including private and corporate clubhouses, retail facilities, food and beverage outlets, banks and entertainment spaces.

The site is surrounded by residential and supertall high-rises. After rounds of modeling and simulation of surrounding cityscape and volumes, the resulting master plan is rational and efficient, allowing for maximum open views. Built around the concept of an oasis, the master plan builds links between built-up areas and the surrounding greenery, integrating the three building plots with the nearby park to meet the official requirement for 40% green space.

The facades of the five supertall high-rises, which range in heights from 160 metres to 220 metres, deliver an architectural statement that combines both organic and contemporary design. Considering the complex yet unique building forms and strict official requirements on spatial efficiency, both the Revit and Rhino architectural programmes were adopted during the design and construction phases.

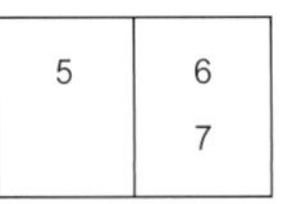

5. 大望京综合开发项目全景
Overview of the project
6. 总平面图
General Plan
7. 多功能商业会展大楼外观
Aerial view of the multi-functional commercial exhibition complex

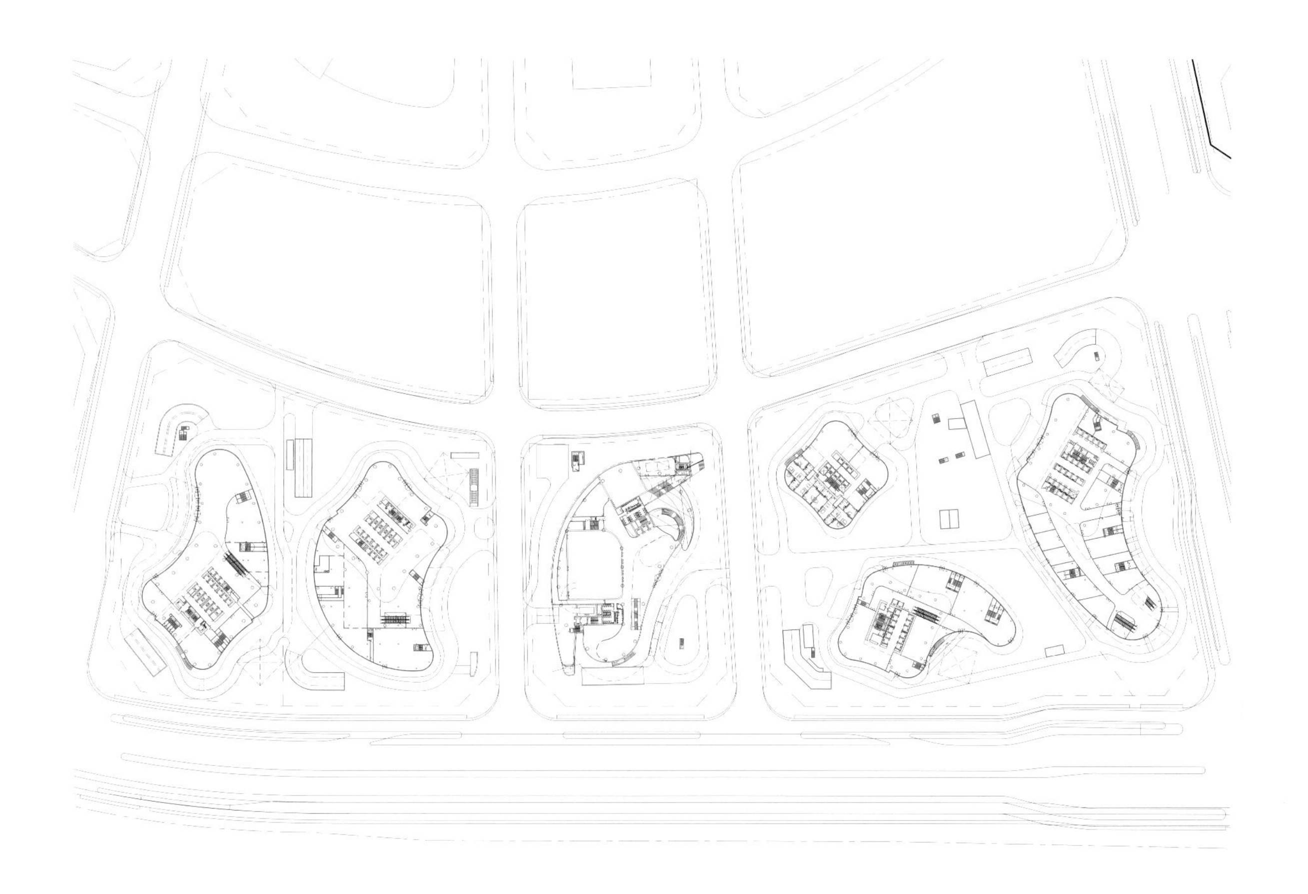

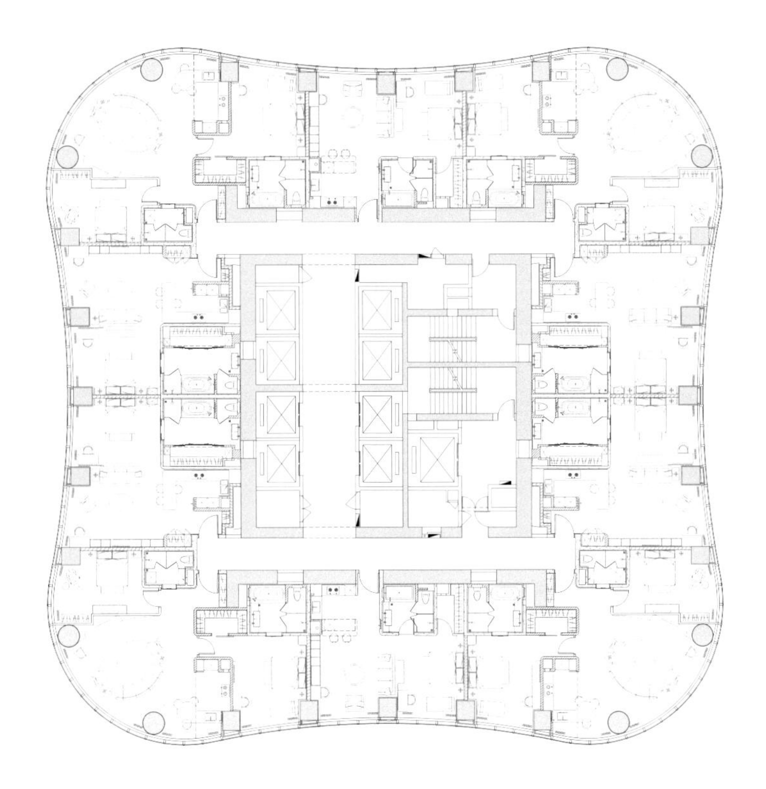

8	9
	10

8. 建筑外观
External view
9. 公寓平面图
Floor plan of residential building
10. 办公楼平面图
Floor plan of office building

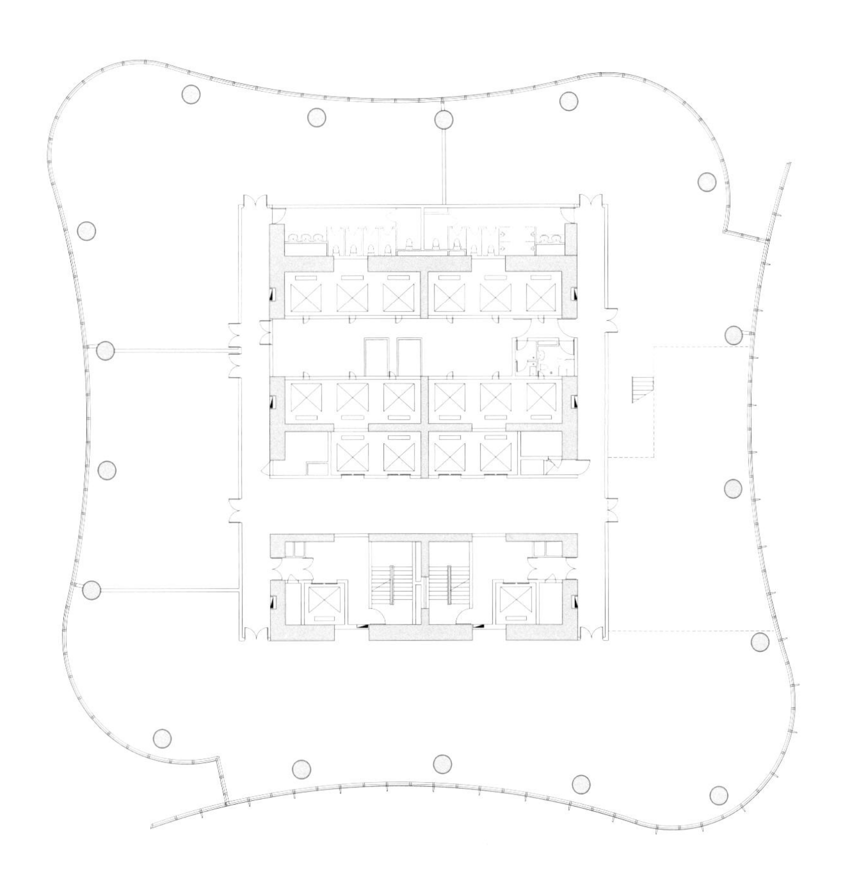

北京大兴 3 及 4 地块项目

Daxing Plots 3 and 4, Beijing

北京大兴 3 及 4 地块项目位于北京东南部的五环路，包括 150 000m^2 的商业零售、180 000m^2 的办公及 35 000m^2 的酒店。基地北侧为宜家（IKEA），南侧为地铁文化公园。项目虽占据了两个地块，但一致的设计语言确保了整个项目的和谐统一。

建筑呈南北向条状，有如山脉起伏的地质形态。“山体”南侧部分降低了体块高度，使得建筑组合更显自然；“山体”北部彰显出更为鲜明的都市形象；“山谷”延伸至南部的景观地带，尊重自然并与文化公园充分融合。

从宜家汇入的主要人流进入商场带动起商业流线。设计减低了商业部分过大的体量，从而获得更合理的商场运营模式及空间。设计还进一步改变总体体量分配以增加办公楼空间，并将主力百货放于商场两端以带动整体商场人流。设计还注重营建商店的临街立面，并将酒店置于商场之上，以拥有更好的景观。

项目以其尊重自然环境的布局态度、灵动轻盈的建筑姿态，带动了整个地区的商业与文化生活活力。

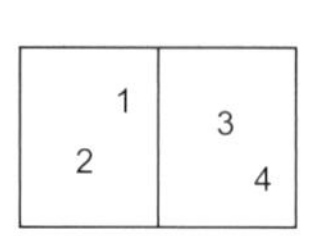

1. 北京城市景观
Cityscape of Beijing
2. 面对公园景观的轻盈立面
View from nearby park
3. 动感、活力的沿街立面
View from vibrant street
4. 项目区位图
Location map

项目信息 | PROJECT INFORMATION

地　　点：中国北京
建筑面积：363 049m^2
设计时间：2011—2015 年
建成时间：2018 年
项目功能：综合体
业　　主：北京兴创置地房地产开发有限公司
主要设计人：Andrew BROMBERG

Location: Beijing, PRC
Gross floor area: 363,049 m^2
Design time: 2011-2015
Completion year: 2018
Sector: Mixed-use
Client: Beijing Xing Chuang Zhidi Real Estate Development Co.,Ltd.
Design Director: Andrew BROMBERG

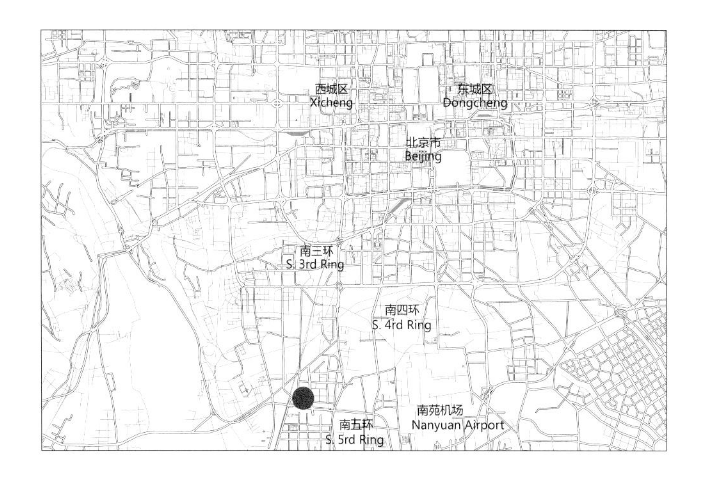

Located in southeast Beijing just off the Fifth Ring Road, Daxing Plots 3 and 4 is a mixed-use development with 150,000 square metres of retail, 180,000 square metres of office space and 35,000 square metres of hotel space. The project takes up two large plots between IKEA to the north and the Metro Culture Park to the south.

A shared design language unifies the development. The architectural design is north-south striated, just like the geological form of rolling mountains. The south side of the "mountain" is lower in height, which makes the massing more natural; its north side emphasises a more conceptual kind of urban architecture. In addition, with respect for nature, the valley between the two masses becomes a landscape belt that extends all the way to the south and is fully integrated into the Culture Park.

The design intends to bring foot traffic from IKEA to the project's shopping mall. Breaking down the oversized commercial volume creates a stronger sense of space that leads to a more open experience. The design further adjusts the building massing to allow more space for the office buildings, and at the same time it locates department stores at both ends of the shopping mall to activate circulation. Prominent shopfronts are placed along the street, while the hotel sits above the shopping mall to maximise its views.

Together with respect for the natural environment, the master plan and architectural language of the project generate commercial and cultural vitality for the entire district around it.

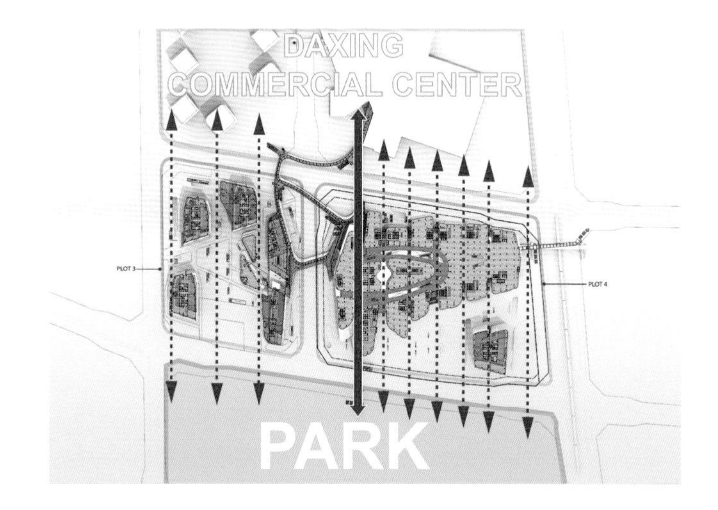

5	7
6	8

5. 鸟瞰
Aerial view
6. 分析图
Analysis diagram
7. 从附近公园视角看项目
View from nearby park
8. 入口大厅
Lobby

北京新浪总部大楼

Sina Plaza, Beijing

城市空间认知

北京，中华人民共和国首都、直辖市和国家中心城市，政治文化中心和经济决策管理中心，是中国“八大古都”之一，蕴含着灿烂的中华文化和中国智慧。北京的传统建筑整体宏伟大气、轴线明确，神秘的空间层层递进展开——正如故宫的层层矩形庭院的推进。新浪总部大楼地处北京西北部的科技总部园区，在规整的基地内，设计采用中国的建筑智慧来创造世界的语言，呈现出中国高速发展的高科技 IT 公司的理念，以简洁的类矩形的建筑体块、轻盈的金属与玻璃立面，很好地诠释出互联网公司的精神：大道至简，由内而生，无限自由。

大道至简：中国智慧

中国崇尚的宇宙之道、自然之道是大道至简，一生二、二生三，而后生无穷。传统的北京城也是由一个个矩形庭院建筑重重叠加而成。新浪总部大楼就设计成这样一个巨大的类矩形平面，大气简洁，内部包罗万象，处处蕴含生机。中国传统观念“天圆地方”，在曲直之间概括了全部的世界空间，如此直观简洁的观念，影响着我们的城市、建筑和生活的空间。新浪总部大楼应合了中国的建筑智慧和观念，以类矩形的建筑，用最简洁的设计手法，让曲直之间相生相通，建筑空间浑然天成，

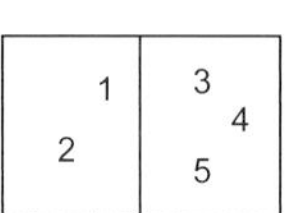

1. 新浪总部的造型如同紫禁城角楼一般平整方正、严谨简洁，亦如北京这座城市的气息
The form of Sina Plaza adopts an architectural language reminiscent of the Forbidden City: rectangular, dignified and simple, like the layout of Beijing itself
2. 从西北向看新浪总部大楼
View of Sina Plaza from the northwest
3. 新浪总部大楼立面
View from street
4. 项目区位图
Location map
5. 设计灵感及手绘图
Design concept and sketch

项目信息 | PROJECT INFORMATION

地　　点：中国北京
建筑面积：124 500m²
设计时间：2012—2014 年
建成时间：2015 年
项目功能：办公楼
业　　主：富力地产开发有限公司、新浪网技术（中国）有限公司
主要设计人：韦业启

Location: Beijing, PRC
Gross floor area: 124,500 m²
Design time: 2012-2014
Completion year: 2015
Sector: Office
Client: R&F Properties, Sina Technology (China) Co., Ltd.
Director: Ken WAI

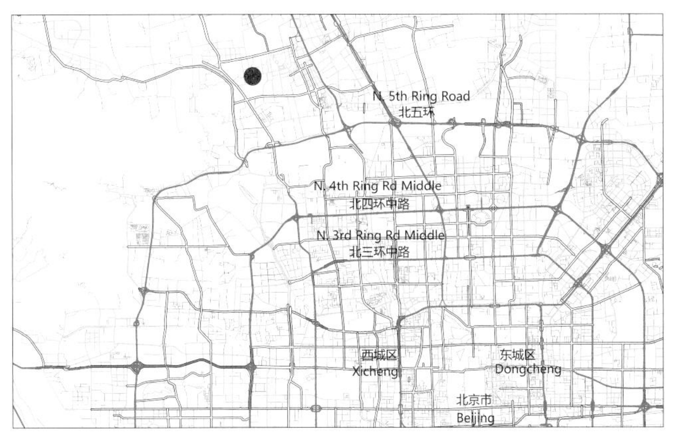

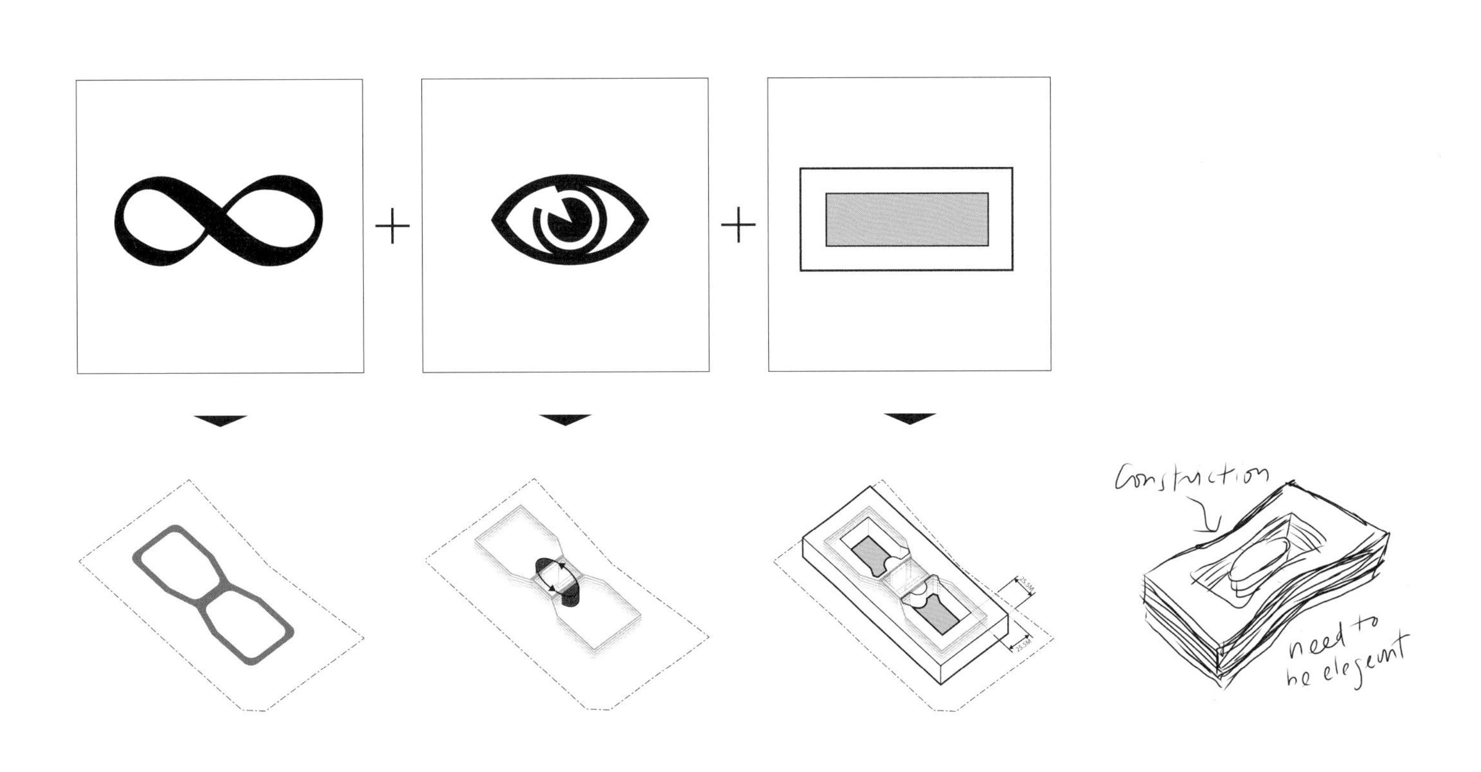

建筑似乎在自然之力的作用下呼吸运动。建筑立面微微隆起，渐渐收进，这种缓慢、微小的变化可以带来无限的想象。建筑的立面几乎是均匀竖向等分，横向的遮阳金属百叶无限延展，在蓝天映衬下，建筑消失在碧蓝的背景中，建筑之形与天合一。

由内而生：空间庭院

让“一家人”在一起工作是新浪总部大楼设计的初心。IT 公司的工作需要创造很多交流互动的机会，因此，走道、楼梯厅等交通空间都成为互动场所，大空间包容了各种的行为方式，灵活的空间布局激发大家面对面的交流，提供了面向未来伸缩自如的空间弹性。巨大的建筑空间需要拥有高性能的空调过滤系统，提升空气的质量，提高环境的品质。新浪大楼的空间基于通过各种调研采集的数据，经分析比较得以明确。让公司员工获得家的感觉，这也是大楼的设计目标之一，因此整栋建筑有行走步道、瑜伽活动、

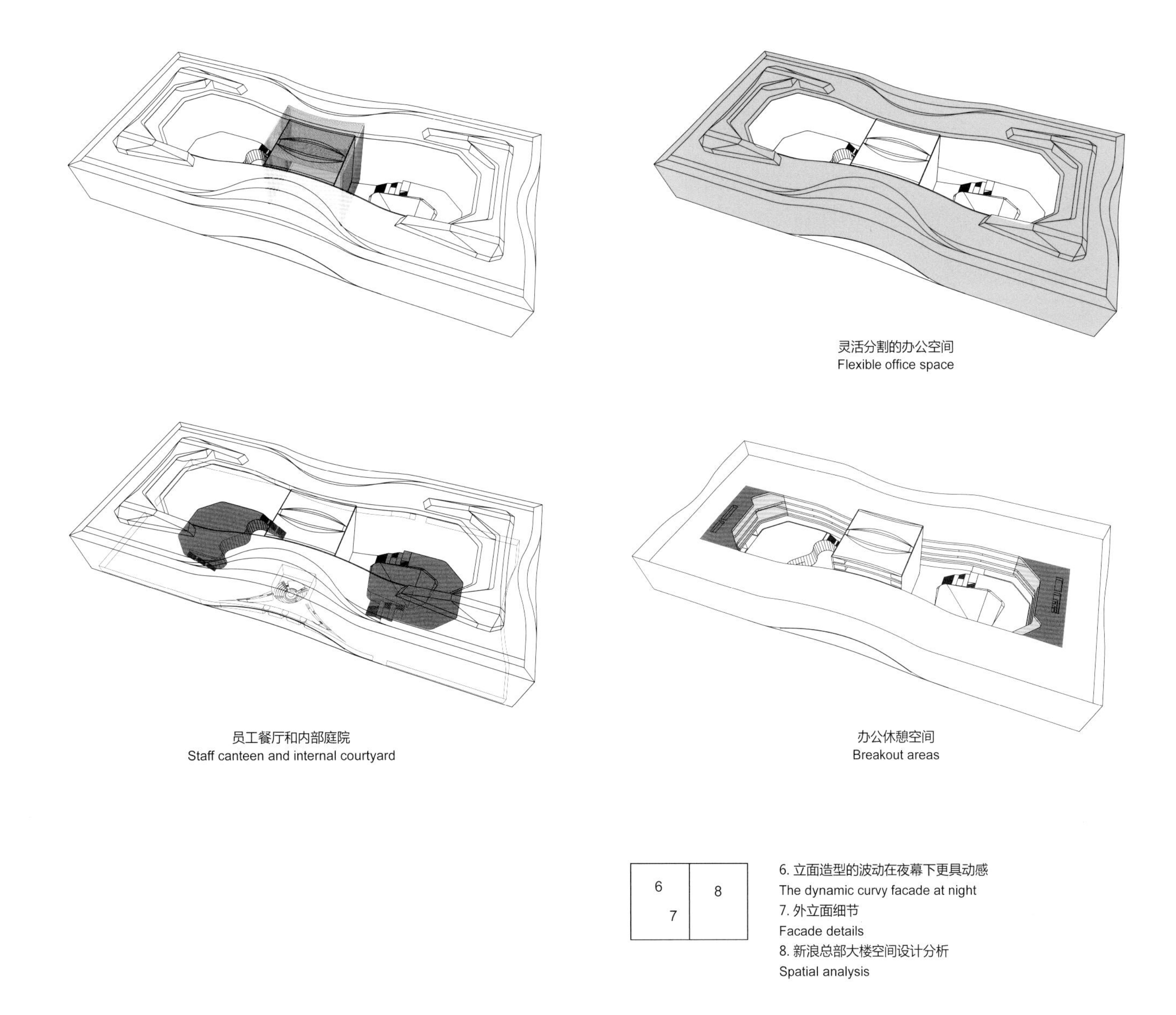

6. 立面造型的波动在夜幕下更具动感
The dynamic curvy facade at night
7. 外立面细节
Facade details
8. 新浪总部大楼空间设计分析
Spatial analysis

游戏、健身、桌球等场地空间，不在办公座位上时可以拿着电脑、iPad 游走于建筑的不同空间，或坐在舒适的沙发上畅想和发呆……建筑留出了两个庭院，塑造出内向的空间，为更多丰富的活动提供场所。内部中庭的“新浪眼”成为建筑空间中的点睛之笔，全球的资讯在此同步互动，新浪与世界永远在转动。

无限自由：变化与人

IT 员工不希望座位是永久固定的，大家在灵活的空间中讨论交流，自由组合的方式能够激发创造力，因为无定形，所以有无穷变化的可能。新浪总部大楼以极简包罗万象，以无形胜有形。建筑以纯净的金属玻璃盒子呈现平静，同时以微妙的形态空间的变化展示建筑空间的张力。这既是 IT 企业的精神，即简洁、透明、纯粹、极致的使命，也是对社会服务平等互联的回应。通透的空间似无穷无尽，这是人类对科技的追求，对无穷尽未来可能性的追求。

About The City

Beijing is the heart of China. As the capital of several grand dynasties, the city is home to a brilliant legacy of Chinese culture and wisdom. Traditional Beijing architecture is characterised by a sense of balance in a dignified and monumental style, with a simple and clear form. A progressive layering of spatial structure intrigues visitors, drawing them into interior spaces. The planning and layout of structures is usually rectangular, as with the courtyards of the Forbidden City. Sina Plaza, the headquarters of the fast-growing technology giant Sina Corporation, is located in the Zhongguancun Science Park in northwest Beijing. The architectural planning and design expressed Chinese wisdom through a global language. It is an approach that embodies the philosophy of the IT companies that make their home here. The simple rectangular form and light metal-and-glass facade are an apt expression of these companies' spirit:

9	10

9. 新浪总部大楼内部中式庭院
Internal courtyard
10. 交通路线的合理设计，使得大楼内任何两点之间的步行时间都不超过 2 分钟
Circulation is carefully planned so that it takes no longer than two minutes to walk from one corner to the opposite one

ambitious yet straightforward, with a powerful core and unlimited freedom.

Grand yet Simple: Chinese Wisdom

Chinese philosophy revolves around the Tao of the Universe and Nature: simple yet powerful. One begets two, two begets three, and so on and so on. Traditional Beijing was a city built on blocks of courtyards and buildings, which are reflected in Sina Plaza's rectangular layout. A variety of lively programmes animate the building's interior. In ancient times, Chinese saw the heaven as "round" and the earth as "square." The straight and curved lines define all spaces in the world. This intuitive and simple concept has shaped our cities, architecture and living spaces. The design of Sina Plaza embodies Chinese wisdom and philosophy. The simple rectangular building hosts a dialogue between straight and curved lines. Spaces are formed naturally, which allows the architecture to "breathe" and operate in an organic way. The facade is almost evenly divided by horizontal and vertical elements. They appear to rise and retreat — the kind of small, subtle variation that ignites the imagination. The horizontal sun-shading louvers shine on sunny days, blending into the blue sky.

Powerful Core: Spacious Courtyard

This headquarters building is designed to house one big family. As the staff of this IT company need ample opportunities to interact, the hallways and staircases are all designed as interactive spaces. The

large areas embrace different types of behaviour, and the flexible spaces encourage face-to-face communications, while also giving the company an opportunity to modify the space in the future. Large architectural spaces such as these require a top-of-the-line air conditioning and filtration system to enhance the quality of air and the interior environment. The aim of the design is for employees to feel at home, which led to the analysis and comparison of design options. The result is a series of spaces for yoga, games, fitness, snooker and more. Employees are not confined to their desks. They can pick up their laptops or tablets and walk around the building to find a table for a lively discussion or a sofa for some quiet time. The two inner courtyards give the building and its users room for different activities. The "Eye of Sina" in the atrium, with a large screen displaying news around the world, connects Sina with the outside world.

Unlimited Freedom: Dancing Through Changes

IT staff do not want to be stuck at the same desk. They interact in flexible spaces that stimulate creativity. Sina Plaza's design is minimalist and thereby inclusive; a free flowing form allows for a diversity of activities. The glass-and-metal architecture embodies transparency and serenity, while the subtle variation between spaces instills a sense of energy. This is what IT companies are about: simplicity, transparency, purity and a return to basics, echoing the egalitarian atmosphere of social media. The transparency of the spaces gives them a sense of endless possibility, just like human beings chasing the advancements of technology to explore a limitless future.

11	12 13 14

11. 宽敞的公共区域用以员工休息和轻松办公
Spacious breakout areas
12. 一层平面图
Level 1 floor plan
13. 北立面图
North elevation
14. 剖面图
Section

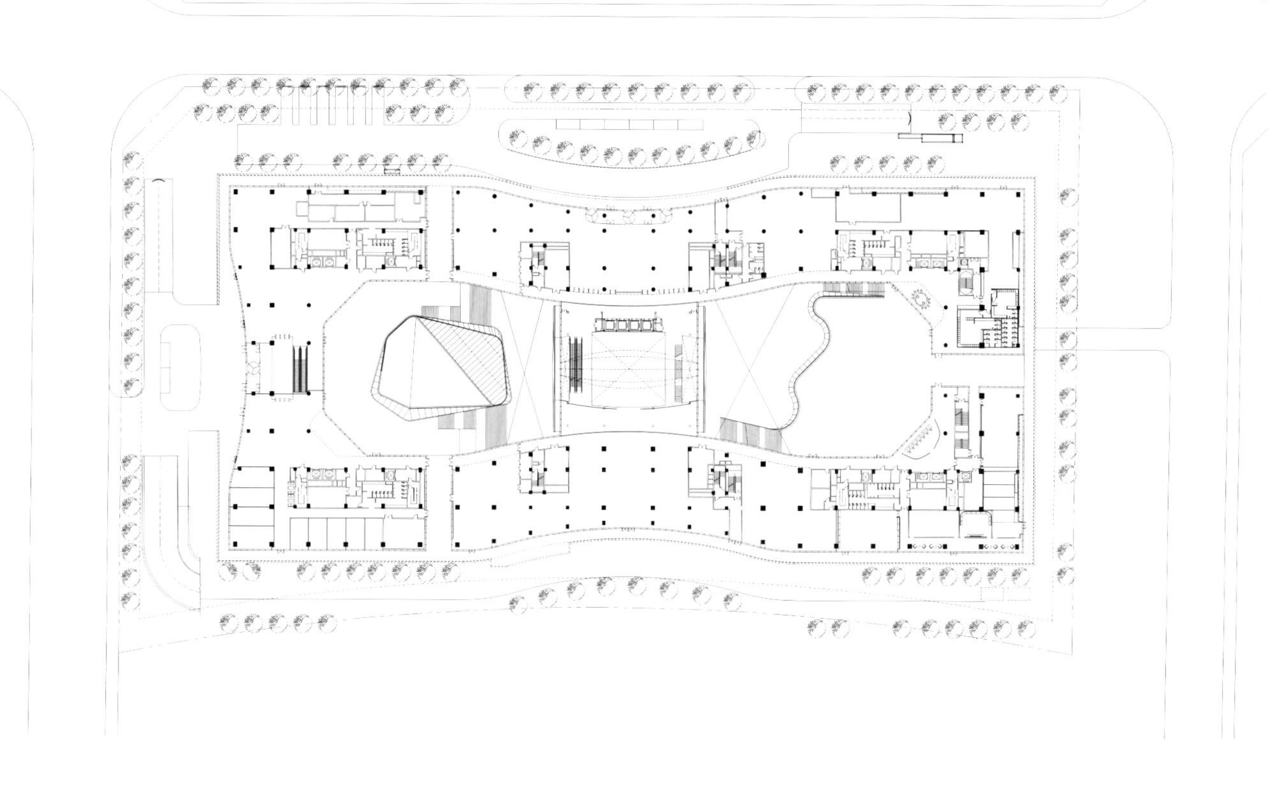

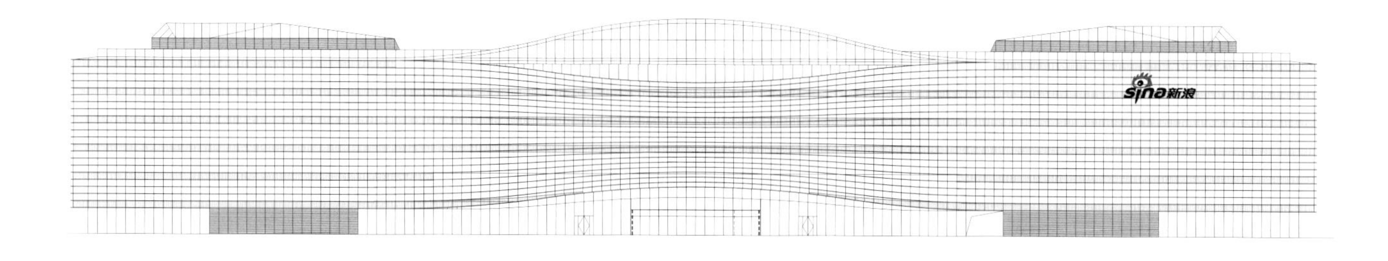
sina新浪

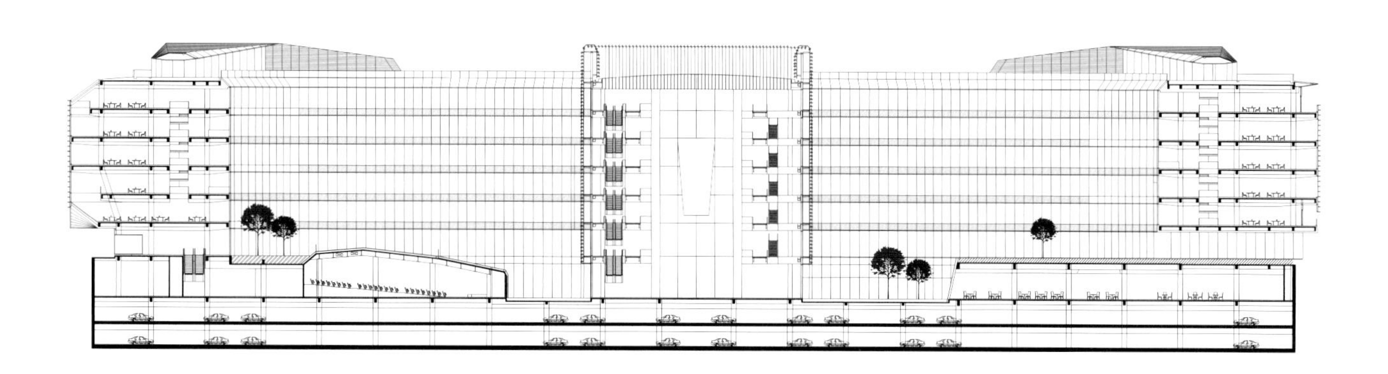

北京北苑北辰综合体

North Star Mixed-use Development, Beijing

远而望之，便惊其势。作为北京奥运会的重要枢纽，北苑北辰综合体开发项目虽在奥运之前及时竣工，但当它以高逾百米的“身姿”携手极具创意的裙楼矗立在这座城市成为焦点时，后奥运时代的活力被彻底激活。项目以“水”为设计理念，穹顶形玻璃屋面扶摇直上，直达裙楼顶端，将其一分为二，随后又如同瀑布般在两边的玻璃“峭壁”之间倾泻而下，形成了瀑布下方人群来往穿梭的入口大门。

玻璃无疑是整座建筑的主角。塔楼外立面采用低辐射镀膜绝缘玻璃，裙楼部分采用了带反射镀膜的夹层玻璃以降低日照热量，在北京这样经常受空气污染的地区，玻璃比其他建筑表皮材质更易于清污。鉴于项目北侧规划为地铁枢纽站，为了更好地将人流引入购物广场，建筑师在地铁站与购物广场中设置了入口，行人可以通过地铁站直接进入购物广场。项目设有包括位于玻璃“瀑布”下方的南入口在内的三个出入口，建筑室内外空间界面形成了和谐的波状弧线。曲面结构的天窗，让光线得以在共享的中庭空间自然流动。

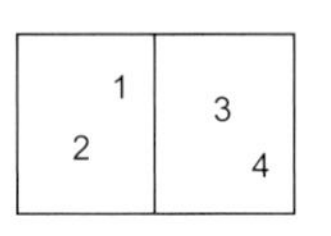

1. 北京古建筑
Traditional buildings in Beijing
2. 3. 建筑外观
External view
4. 项目区位图
Location map

Photography © Paul Warchol

项目信息 | PROJECT INFORMATION

地　　点：中国北京
建筑面积：161 780m²
设计时间：2004—2008 年
建成时间：2010 年
项目功能：综合体
业　　主：北京北辰实业股份有限公司北辰置地分公司
主要设计人：Andrew BROMBERG

Location: Beijing, PRC
Gross floor area: 161,780 m²
Design time: 2004-2008
Completion year: 2010
Sector: Mixed-use
Client: Beijing North Star Company Ltd.
Design Director: Andrew BROMBERG

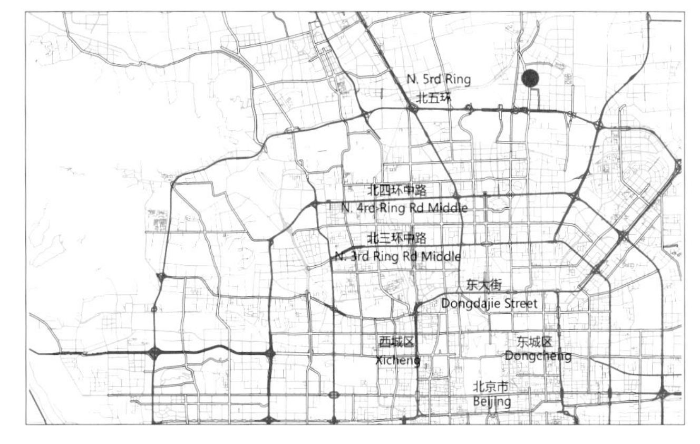

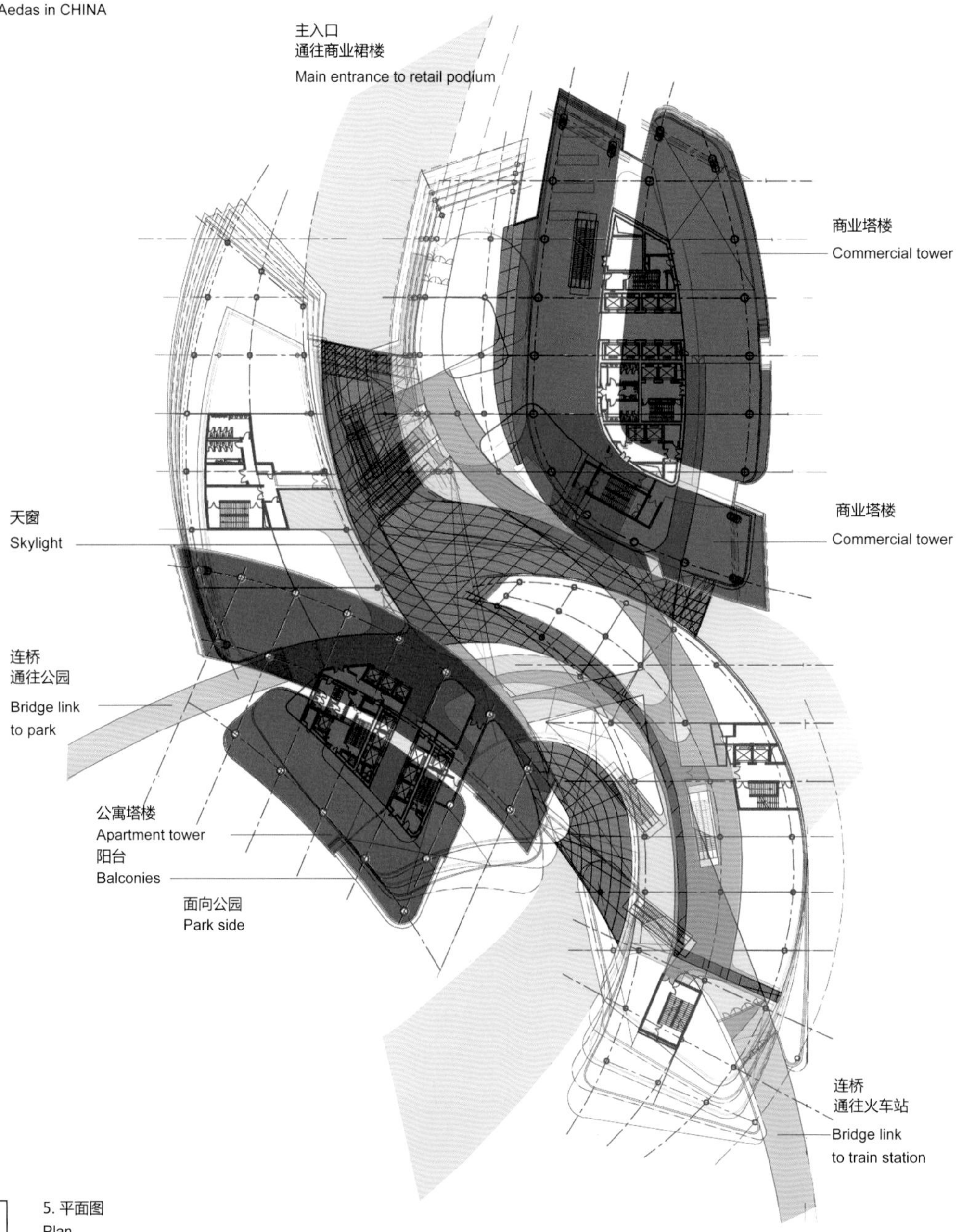

主入口
通往商业裙楼
Main entrance to retail podium

商业塔楼
Commercial tower

商业塔楼
Commercial tower

天窗
Skylight

连桥
通往公园
Bridge link
to park

公寓塔楼
Apartment tower
阳台
Balconies

面向公园
Park side

连桥
通往火车站
Bridge link
to train station

5. 平面图
Plan
6. 内部中庭
Atrium

This building is striking in scale, catching your eye even from far away. The North Star Mixed-use Development in Beijing was originally intended to be one of the hubs for the Beijing Olympics. It did make it in time, but some years later its two towers, more than 100 metres high, flanked by an innovative six-storey shopping podium were back in the city's spotlight. The project drew inspiration from water; the twisted glass surface rises steeply to the middle of the towers and slices the volume into two, and then pours down like waterfall between the glass cliffs and forms the main entrance of the building.

The glass form is undoubtedly the protagonist of the building. The tower facades are in insulated glass with Low-E coating and the podium's laminated glass has a reflective coating used to reduce solar heat gain. This allows Beijing's pollutants to be easily cleaned from glass. A subway station was slated to be built on the north side of the development and one of the ambitions of the design was that people would come out of the station and go straight into the mall through a specially designed exit/entrance. There are three entrances planned, including the south entrance under the metaphorical waterfall. The effect is that people can simply drift from the south side into the shopping environment and find its flowing curves and undulations not too different from the qualities they enjoyed in the outdoor space. The snaking of the geodesic skylight across the roof of the podium allows light to pour through and flow in the atrium.

Photography © Paul Warchol

青岛金茂湾购物中心

Jinmao Harbour Shopping Center, Qingdao

金茂湾购物中心位于青岛这座享有“东方瑞士”美誉的港口城市的西海岸线，胶州湾蓝海新港城综合体项目的核心位置。该购物中心包含地上 5 层商业零售和地下 2 层停车场。通过室外阶梯式花园引导人流进入码头上方的空中公园和海滨长廊，并与购物中心及周边开发项目相连接。

建筑外形设计灵感来自青岛海岸线上层层交错的页岩。为呼应自然地貌，各楼层相互堆叠，形成高低不平的页岩状外观。设计又从集装箱海运物流中汲取灵感，通过组合耐候钢、扩张网、木材、砖块、着色混凝土覆层，保留住城市的工业质感，追溯那段令人印象深刻的沿海货运、仓储业蓬勃发展的工业时代。

1. 青岛是中国沿海重要的中心城市，拥有世界最大港口之一，金茂湾购物中心的设计从青岛海岸线特殊的地貌及集装箱运货船汲取了灵感
Qingdao is a coastal city with one of the world's biggest ports. The design drew inspiration from the unique landscape of Qingdao and container ships.
2. 金茂湾购物中心外景
External view
3. 极具工业气息的材质保留了城市的原有质感
Use of materials that reflect the character of the city
4. 项目区位图
Location map

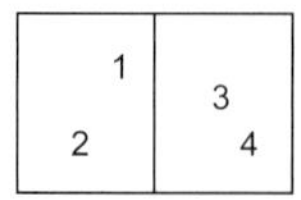

项目信息 ｜ PROJECT INFORMATION

地　　点：中国青岛
建筑面积：34 530m²
设计时间：2013—2017 年
建成时间：2017 年
项目功能：商业零售
业　　主：方兴地产（中国）有限公司
主要设计人：纪达夫

Location: Qingdao, PRC
Gross floor area: 34,530 m²
Design time: 2013-2017
Completion year: 2017
Sector: Retail
Client: Franshion Properties (China) Limited
Director: Keith GRIFFITHS

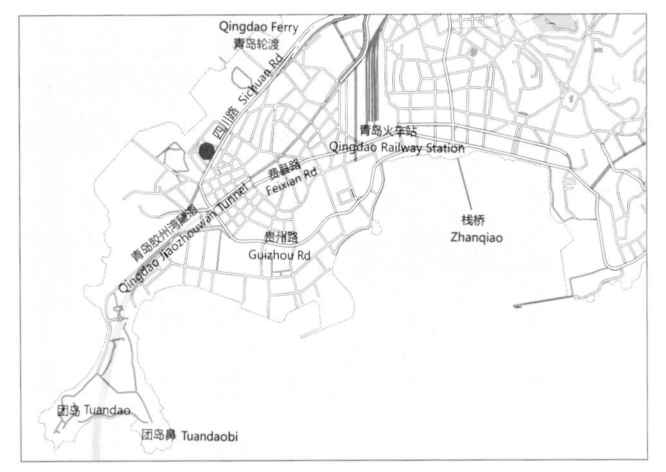

The Jinmao Harbour Shopping Center is located in the core of the Lanhai New Harbour City on Jiaozhou Bay, the western shoreline of Qingdao, which is a city with the reputation of being the "Oriental Switzerland." The mall offers five levels of above-ground retail and two levels of basement parking. An outdoor stepped garden drives circulation to the sky-park above the wharf and the seaside promenade, and connects the adjacent developments.

Along the shoreline of Qingdao, strong waves have crafted a natural landscape of layered shale staggered on top of each other. The staggered shale is transferred into the architectural form of the mall as an homage to the natural landscape. The floors are stacked on top of each other to form a rugged shale-like exterior that creates an impressive industrial reminiscence of the freight industry which flourished along the coast.

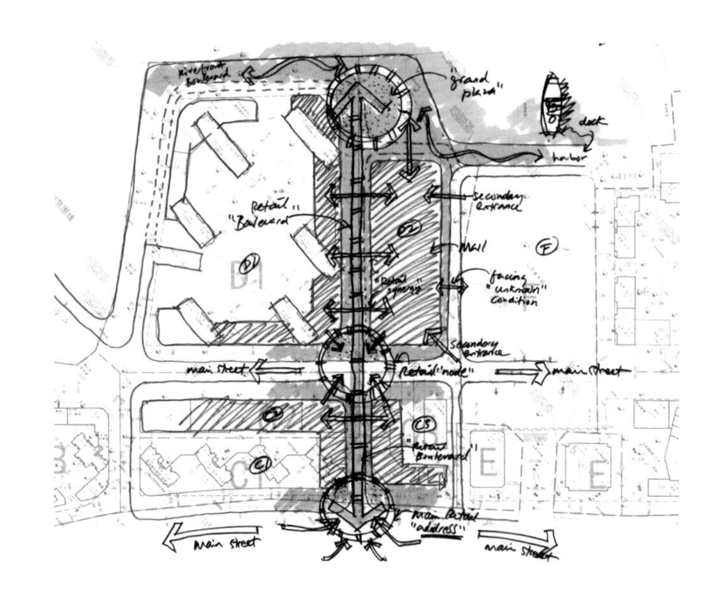

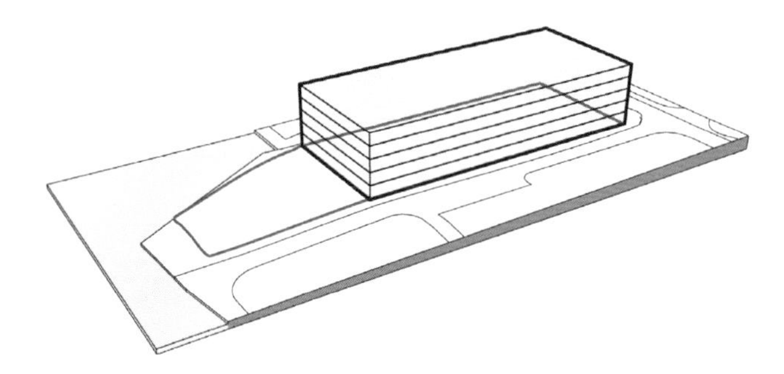

步骤一：体量堆叠
项目建筑面积计算。通过堆叠简洁的体量来满足项目的要求。

STEP 1 MASSING STACKING
Calculation of floor area of the project. Meeting the requirement of the project by stacking simple building volumes.

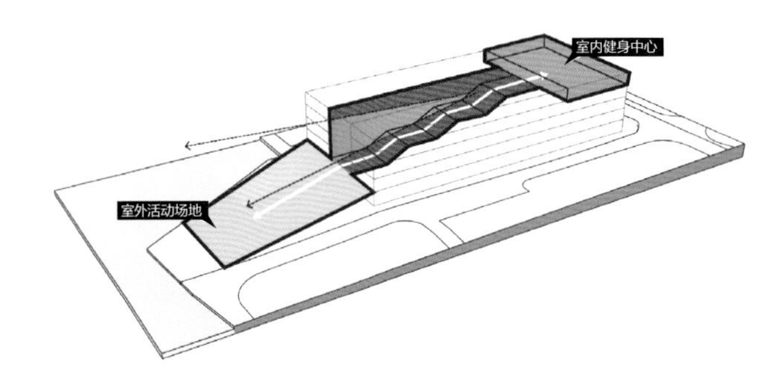

步骤二：室内外连接
设计将购物中心和社区及周围景观融合在一起，从而提供独特的购物环境和社交平台。它将海岸风景和都市生活联系在一起。

STEP 2 LINK INTERIOR AND EXTERIOR
The shopping center is designed to blend into the community and landscape, in order to provide a unique retail and social platform. It links the coast landscape with urban lifestyle.

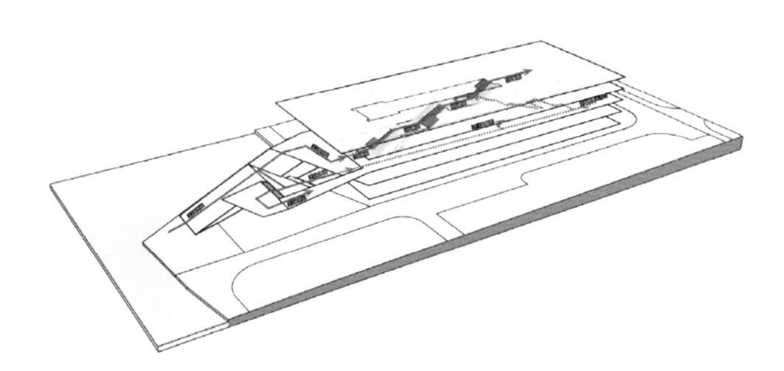

步骤三：优化零售动线
室内主要由中庭及线型动线沟通，直观高效。室外动线与景观紧密结合，打造360°浸入式景观。

STEP 3 OPTIMIZING RETAIL CIRCULATION
The internal circulation is mainly structured through the atrium with a linear circulation, which is direct and efficient. The circulation outside is closely connected with the landscape to create 360°view.

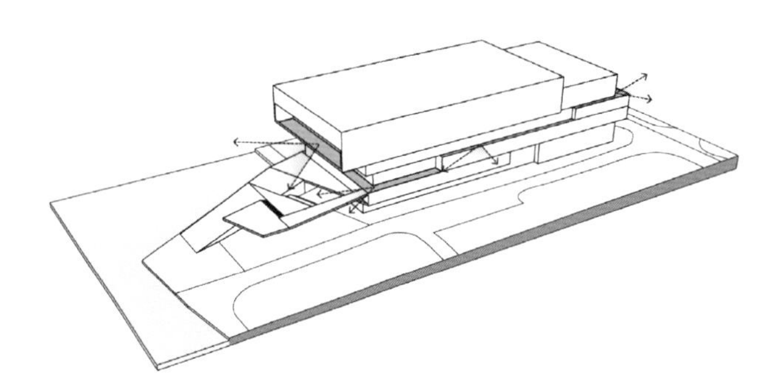

步骤四 室外景观露台
阶梯式花园打造观看海景的露台，躲避冬季寒冷的西北风，庇护绿色景观空间。

STEP 4 OUTDOOR VIEW TERRACES
The stepping garden is designed strategically to create terraces for sea views as well as to diffuse the cold northwestern wind during winter and shield the adjacent green space.

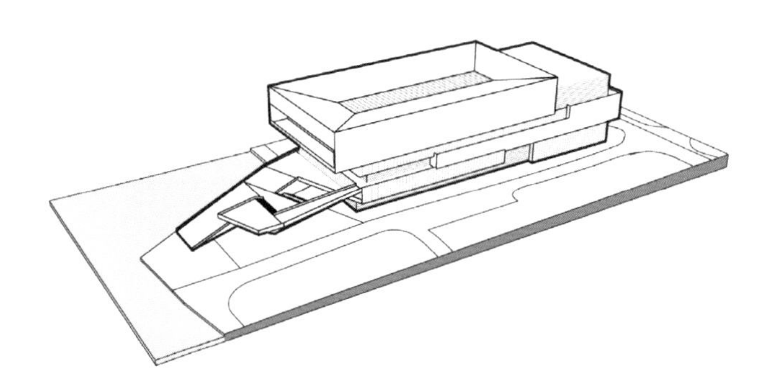

步骤五：自然光照
通过设置屋顶天窗和玻璃立面，该项目利用自然光来减少能耗

STEP 5 NATURAL LIGHT
By creating rooftop skylight and glass facade, the project makes use of the natural light in order to reduce energy consumption.

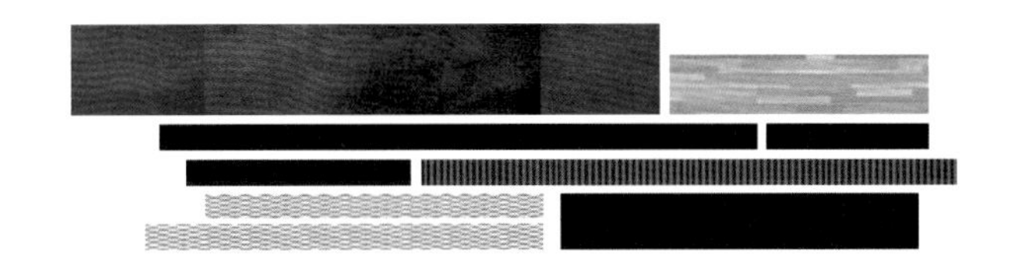

5	7 8
	9 10
6	11 12

5. 金茂湾购物中心内部
Interior view of the shopping mall
6. 手绘图
Sketch
7-11. 空间流线分析图
Analysis diagrams
12. 设计将交错的页岩转化为购物中心的建筑外形，呼应自然地貌
In response to the natural landscape, the design translates the natural layered shale into a building form

星美国际影商城
STELLAR INTERNATIONAL CINEPLEX MALL
4F
衬衫大放价
全场大均一
100元
150元

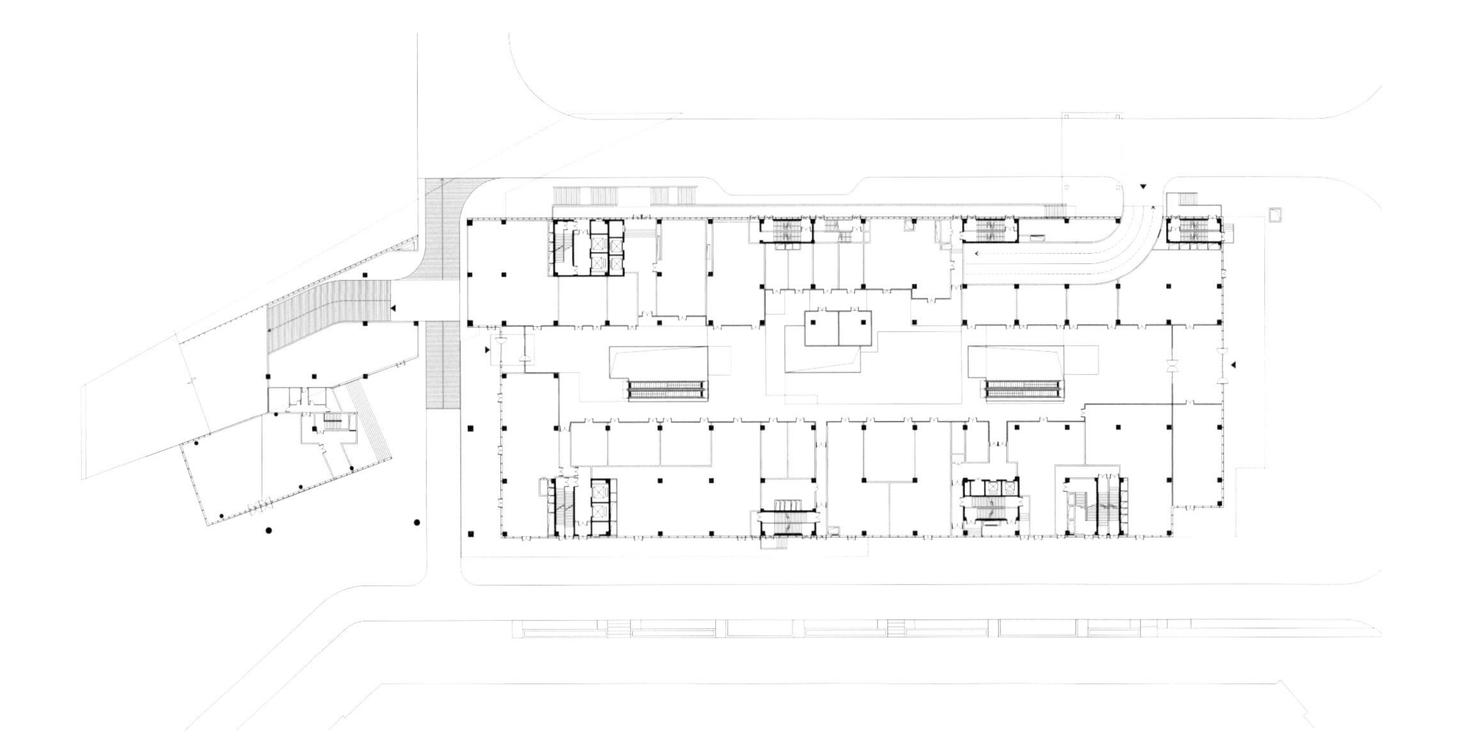

13	14
	15
	16

13. 中庭
Atrium
14. 一层平面图
Level 1 floor plan
15. 北立面图
North elevation
16. 剖面图
Section

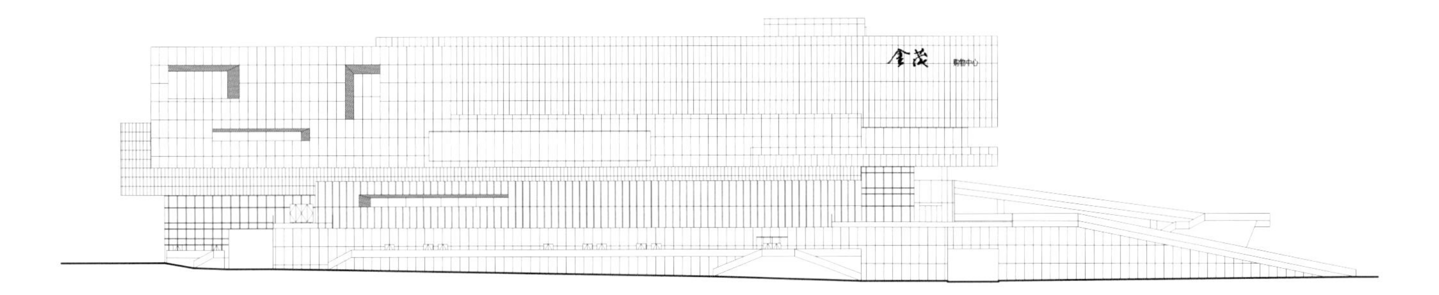

成都恒大广场

Evergrande Plaza, Chengdu

成都，四川省会城市，承载着近三千年的悠久历史，拥有众多名胜古迹和人文景观，平静而祥和地屹立于"天府之国"的腹地。恒大广场作为成都市老城区复兴计划的重点项目，位于青羊区新中央商业区的核心位置，距市中心的天府广场和春熙路仅数步之遥。项目的设计灵感来自四川黄龙风景区美轮美奂的自然地貌景观，该地区堪称天然钙体博物馆，尤其是天然彩池闻名世界，被誉为"人间瑶池"。设计汲取天然梯田水池和瀑布山川的地形特征，呼应成都繁华洒脱、悠闲宁静的城市气质，结合不同功能的空间需求，实现了形式与功能的完美平衡，为充满活力的城市中心创造出一座具有地域特色的都市绿洲。

恒大广场包含 THE ONE 购物中心、六星级成都瑞吉酒店、恒大甲级办公楼及两栋"成都之门"住宅塔楼。建筑打破室内外商业的界限，融合了一系列蜿蜒层叠、不同高度的平台，营造出流动的建筑形态和空间。四幢塔楼拥有卓然不同的立面设计，却又和谐统一，它们围绕购物中心错落分布，在引导人流的同时，保证每层都能获得最开阔的视野。

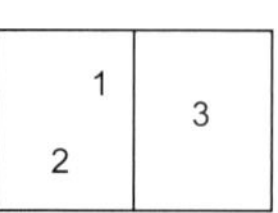

1. 恒大广场"高山流水"般的造型，呼应了成都"天府之国"的繁华洒脱、悠闲宁静的城市气质
Evergrand Plaza, reveals the temperament of Chengdu
2. 恒大华置甲级办公楼在基地西面拔地而起，与东南角的成都瑞吉酒店相视而立，宛如通往室外绿色街道的门户
HD Center and The St. Regis Chengdu
3. 商业零售区入口旁醒目的卵形斜肋构架结构象征着技术的进步，复杂曲面以及 700 片三角玻璃的形态均由先进的参数化软件设计完成
The eye-catching ovoid structure at the southern entrance is achieved with the help of parametric software

STARBUCKS COFFEE
06

项目信息 | PROJECT INFORMATION

地　　点：中国成都
建筑面积：430 000m^2
设计时间：2007—2010 年
建成时间：2015 年
项目功能：综合体
业　　主：恒大地产
主要设计人：林世杰

Location: Chengdu, PRC
Gross floor area: 430,000 m^2
Design time: 2007-2010
Completion year: 2015
Sector: Mixed-use
Client: Evergrande Group
Director: Ed LAM

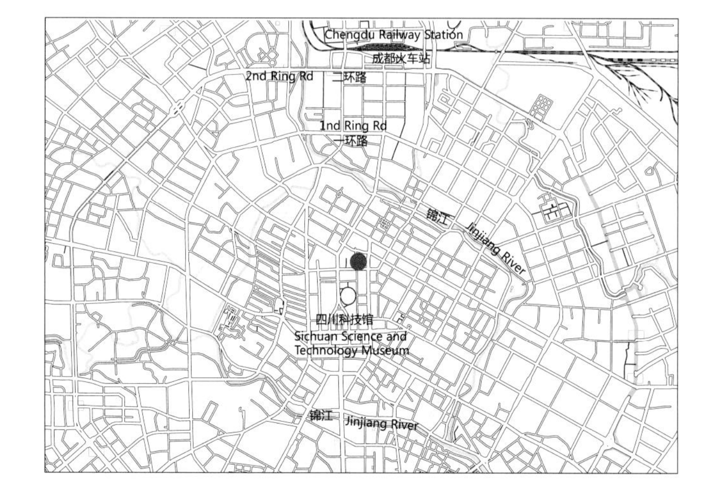

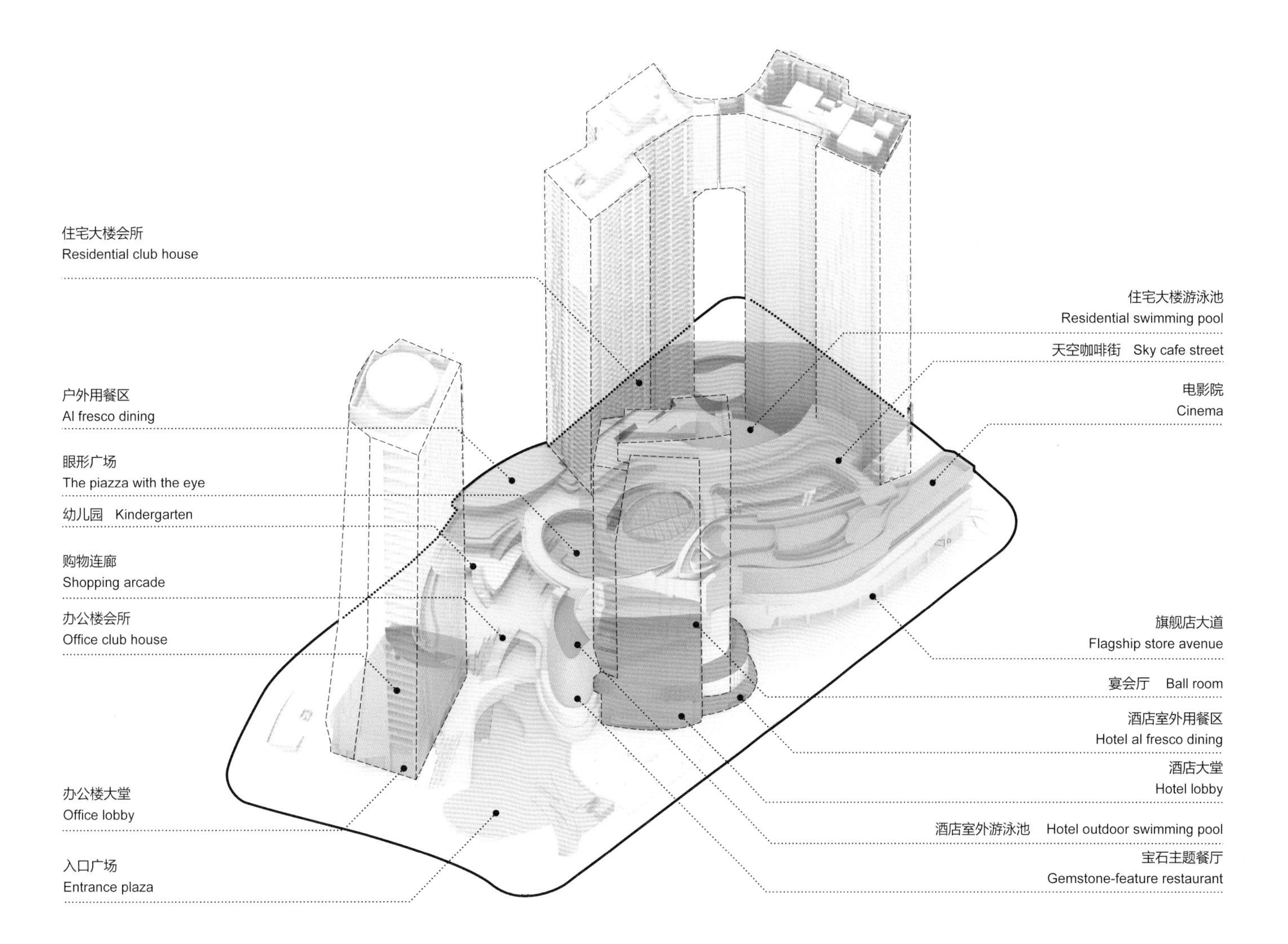

Chengdu, the provincial capital of Sichuan, lies at the heart of the "heavenly land of plenty," as it is known in China. With 3,000 years of history, this city has numerous historic sites and culturally significant landscapes. Evergrande Plaza, located in the core of a new central commercial area in Qingyang district, only a few steps away from downtown and Chunxi Road, is a key project for the revival of Chengdu's historic center. The design is inspired by the natural landscape of Sichuan, in particular the stunning Huanglong Scenic and Historic Interest Area. Huanglong is known for its natural calcium deposits, which have created colourful pools nicknamed "paradise on earth." Evergrande Plaza drew inspirations from the geographical features of these natural terraced pools, waterfalls and mountains to echo Chengdu's temperament of prosperity, freedom and leisure, balance between form and function, and to create an urban oasis within the vibrant city center.

Evergrande plaza consists of THE ONE, the six-star St. Regis Chengdu hotel, the Grade-A office tower HD Center, and two residential buildings named Gate of Chengdu. The green spine of the shopping center blurs the boundaries between the indoor shopping mall and outdoor retail streets. Such spatial design reflects the city's indigenous, outdoor, leisurely lifestyle, with outdoor terraces for al-fresco dining and drinking, alongside an urban park for the community. Four towers with distinctive yet harmonious facades, are distributed around the shopping center. The placement of the towers was chosen based on circulation patterns as well as to take advantage of the best views from the towers.

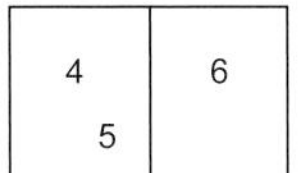

4. 被灯光照亮的建筑综合体入口，下部采用玻璃外立面，上部使用花岗岩覆面
The northern entrance is well lit at night
5. 项目区位图
Location map
6. 恒大广场各部分功能分析图
Functional programmes

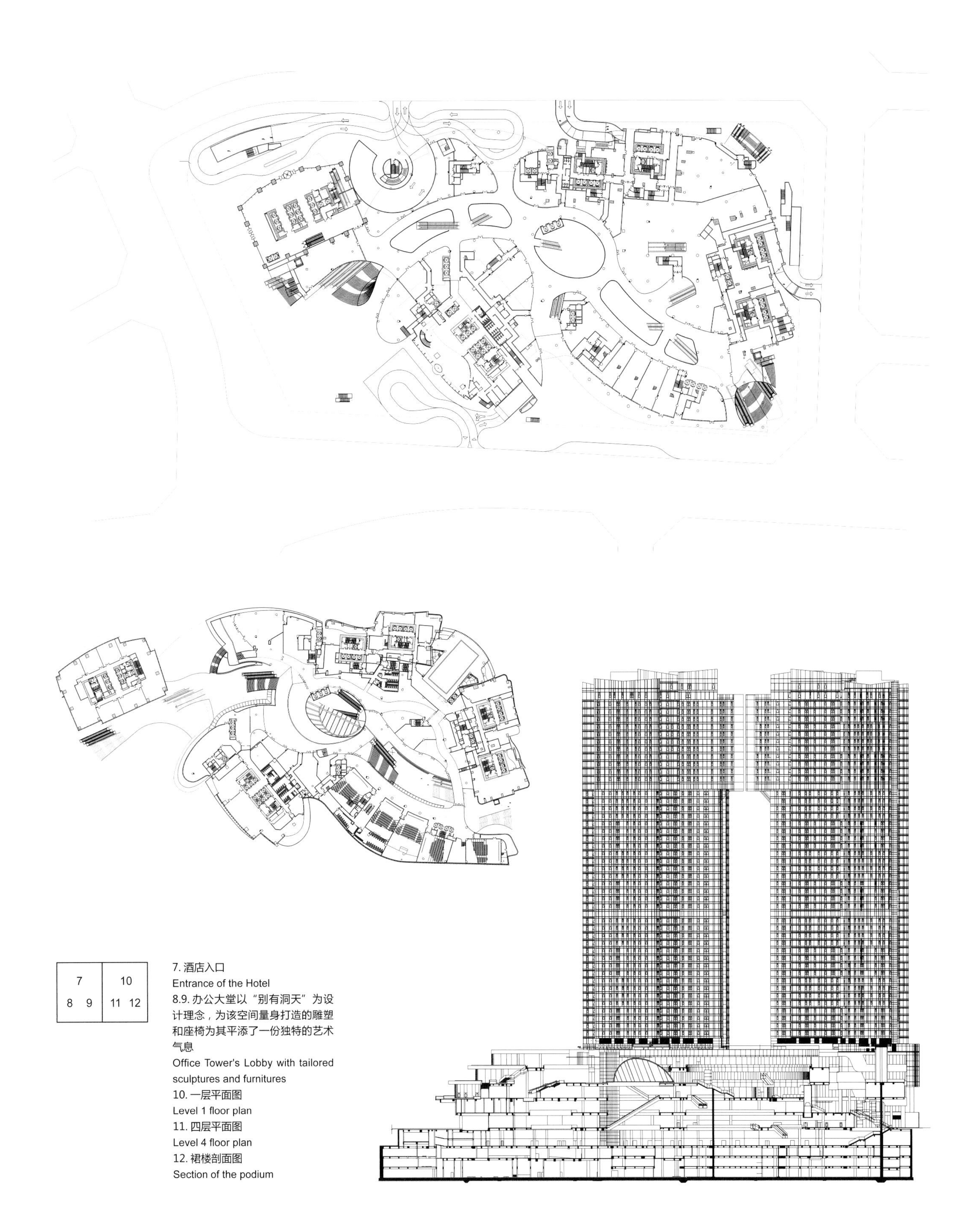

7	10
8 9	11 12

7. 酒店入口
Entrance of the Hotel
8.9. 办公大堂以“别有洞天”为设计理念，为该空间量身打造的雕塑和座椅为其平添了一份独特的艺术气息
Office Tower's Lobby with tailored sculptures and furnitures
10. 一层平面图
Level 1 floor plan
11. 四层平面图
Level 4 floor plan
12. 裙楼剖面图
Section of the podium

重庆新华书店集团公司解放碑时尚文化城

Xinhua Bookstore Group Jiefangbei Book City Mixed-use Project, Chongqing

天梯与梯田：山城意象

山城重庆，地势高低起伏，建筑错落变化。城市中心重庆解放碑 CBD 核心区现有城市密度极高，大型商场林立，商业类型趋同性严重，且商场的外立面过于封闭死板且千篇一律，导致商业核心区没有形成内外的互动关系。如何设计一座体现重庆自身特色，而且能够与解放碑商圈形成互动关系的商业综合体，作为新型地标独树一帜，成为本项目的巨大挑战。Aedas 以“书卷缱绻掩山城”为设计意向，整体建筑一层层延伸向上形成主楼挺拔的天柱，像是一册卷轴，面向城市核心区徐徐展开，一层一层地讲述它的故事。裙楼共 12 层，分为三个部分：1—4 层为商场，5—8 层为书店，9—12 层为餐饮。天梯直达 9 层，建筑裙楼的层层退台形成一系列空中的室外平台，不仅适合山城户外活动的需要，也在拥挤林立的解放碑中心打造出了梯田的自然栖居意向。三座大型空中平台广场，集合了城市活力的各方面，串联起购物、阅读、餐饮、娱乐、休闲、办公等功能，强烈的形态特征与业态的聚合互补，形成了新的商业地标，充满了活力与动感。重庆生活在此叠加，无形的城市精神与有形的建筑空间复合在一起，这是对地形、文脉的理解，是对生活、日常的领悟，是集大成者，是对解放碑核心区的城市空间的注释。

1. 重庆独特的地理环境造就了其不同寻常的城市特征，这些个性鲜明的重庆印象都体现在设计中
The special urban characteristics of Chongqing are attributed to its unique geographical environment. The distinct characteristics are reflected in the design
2. 俯瞰空中生态广场效果图
Aerial view of the sky-plaza
3. 鸟瞰效果图
Aerial view

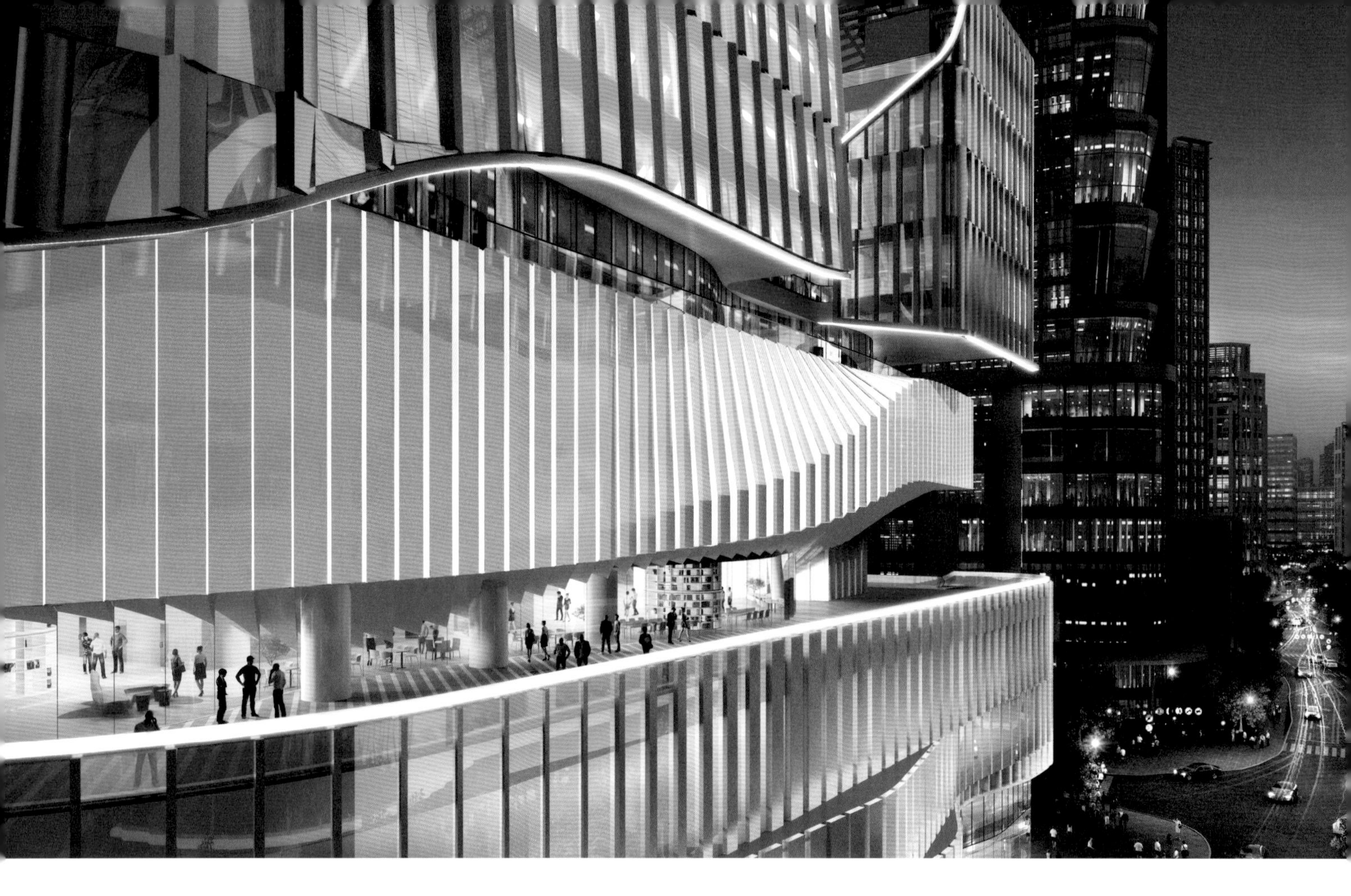

项目信息 | PROJECT INFORMATION

地　　点：中国重庆
建筑面积：153 980m²
设计时间：2014 年至今
建成时间：2020 年
项目功能：综合体
业　　主：重庆新华书店集团公司、重庆新华传媒有限公司和重庆北青实业有限公司
主要设计人：温子先博士

Location: Chongqing, PRC
Gross floor area: 153,980 m²
Design time: 2014 till now
Completion year: 2020
Sector: Mixed-use
Client: Chongqing Xinhua Bookstore Group, Chongqing Xinhua Media Ltd., Chongqing North Green Industry Co., Ltd.
Director: Dr. Andy WEN

金色书屋：建筑意象

内容决定形式，城市书屋即是新华书店的本源。位于建筑 5 层的黄金书屋，是“书卷缱绻掩山城”的点睛之笔。山城空中的一挂金色，如同城市上的明灯，点亮了 CBD，引领大家的目光向上，望向更远的未来。金色书屋已成为城市文化的汇聚地，在 5 层的空中广场开展各种活动，演绎着山城文化，城市居民在此体验着书本蕴含的文化魅力的同时，又能参与到广场的活动中，沉浸在重庆的无限美好中。

空中广场：活力图景

广场是人流汇聚和集散的中心，在重庆解放碑商业中心，项目设计了三个功能不同、海拔不同的广场：地面的商业广场、第 5 层的文化广场和第 9 层的美食广场。在高密度的城市核心区塑造出的天台与峡谷，带来了清新的“空·间”：CBD 中心有了自如的呼吸，人们有了更多可以驻足停留的场地。在城市上空不同高度的广场上，有丰富的活动内容吸引着大家，同时，这些开敞空间又延展渗透到周围的城市街道和空间中，共同融合成解放碑地区的城市活力图景。建筑似从“山”中生长出来，强化出重庆作为山城的特征。

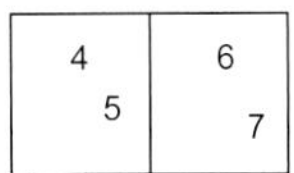

4. 空中文化广场既是新华书店的空中入口，又是向公众开放的文化空间
The sky cultural plaza serves as the entrance to the bookstore and a public cultural space
5. 项目区位图
Location map
6. 概念推导及手绘图
Design concept and sketches
7. 建筑外观人视效果图
External view

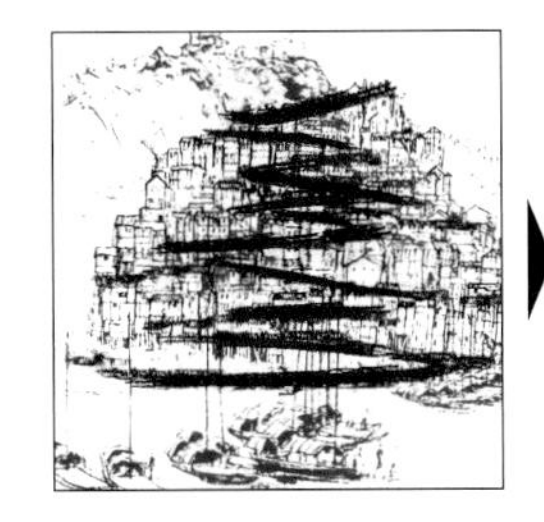
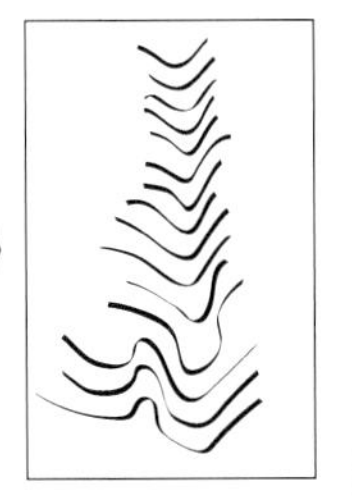

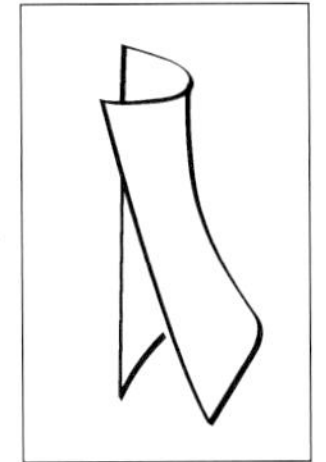

Ladder and Terrace: Working with Mountain City

In Chongqing, the Mountain City, complicated topography brings about buildings that vary in height. In the Jiefangbei central business district, the density is high and there are many large-scale malls, with their facades closed off to the outside and a lack of strong pedestrian connections to outdoor commercial spaces. Rectifying this situation is this project's design challenge. The resulting building unfolds across stepping terraces, with the elegant form of a scroll, implying the spirit of wisdom and knowledge. The stepped architecture not only reveals the well known geographic beauty of Chongqing, but it also interacts with the surroundings and rejuvenates the city landscape. The 12-storey podium is divided into three parts. The first to fourth floors are for retail, the fifth to eighth floors are for the Xinhua Bookstore, and the ninth to 12th floors are for restaurants. The stepped architecture creates a series of terraces. Three large plazas are well connected with the street and green terraces by a grand express escalator and will become a new cultural destination for lifestyle and entertainment activities, enriching and extending the civic space of Jiefangbei by providing a refreshing, tranquil environment for people in this business center to relax and enjoy.

Golden Bookstore: Architectural Intention

Form follows functions. “A city bookstore” is the original idea behind Xinhua Bookstore, something especially true on the golden-hued fifth floor, which shines in the sky of the Mountain City, brightening the central business district and drawing eyes towards it. The golden bookstore will become a civic gathering spot, with many cultural activities that take place in its fifth-floor plaza. City residents can not only enjoy reading in the bookstore but also take part in activities in the plaza.

Sky-plazas: Scene of Vitality

Plazas are urban gathering places and this complex has three distinctively themed plazas. One is a commercial plaza on the ground floor, another is a cultural plaza on the fifth floor. The third is a food plaza on the ninth floor. The project's terraces and canyons add a refreshing texture to the high-density city center, giving it more room to breathe. People have more places to rest and enjoy themselves. Various activities attract them to the plazas on different levels. At the same time, these open spaces extend to and penetrate the surrounding streets and spaces, creating a more vital street life in the Jiefangbei area. The architecture is derived from Chongqing's image as the Mountain City, and it manages to reinforce this character. The building offers a ladder and a terrace in the city, a golden bookstore in the sky, and a plaza in the air.

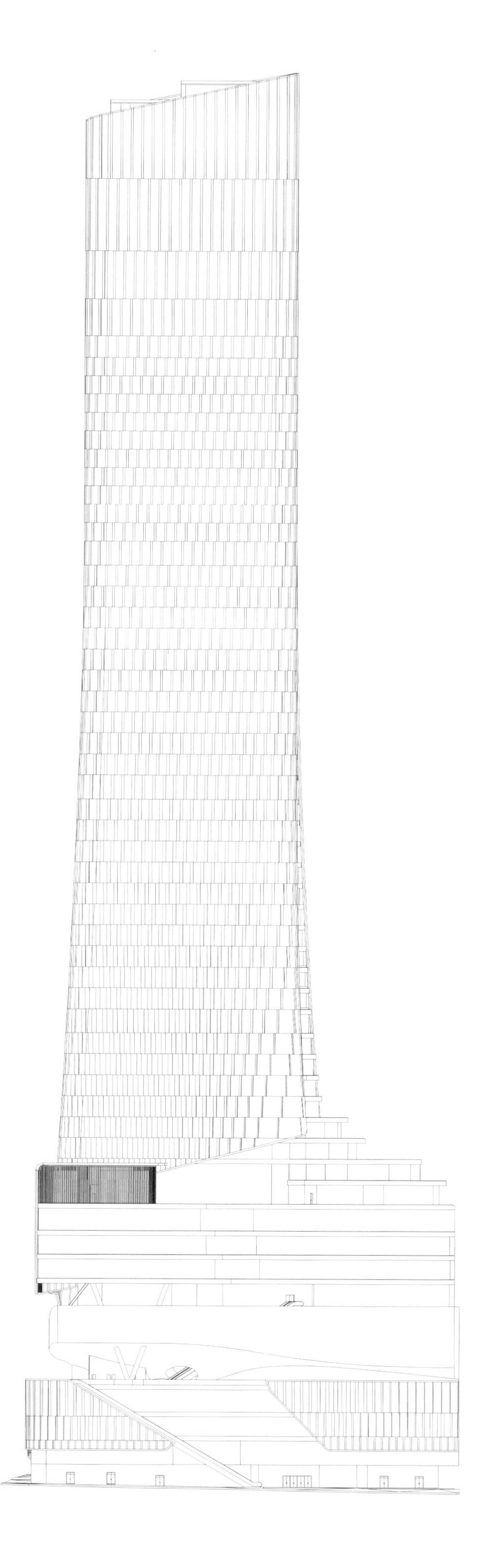

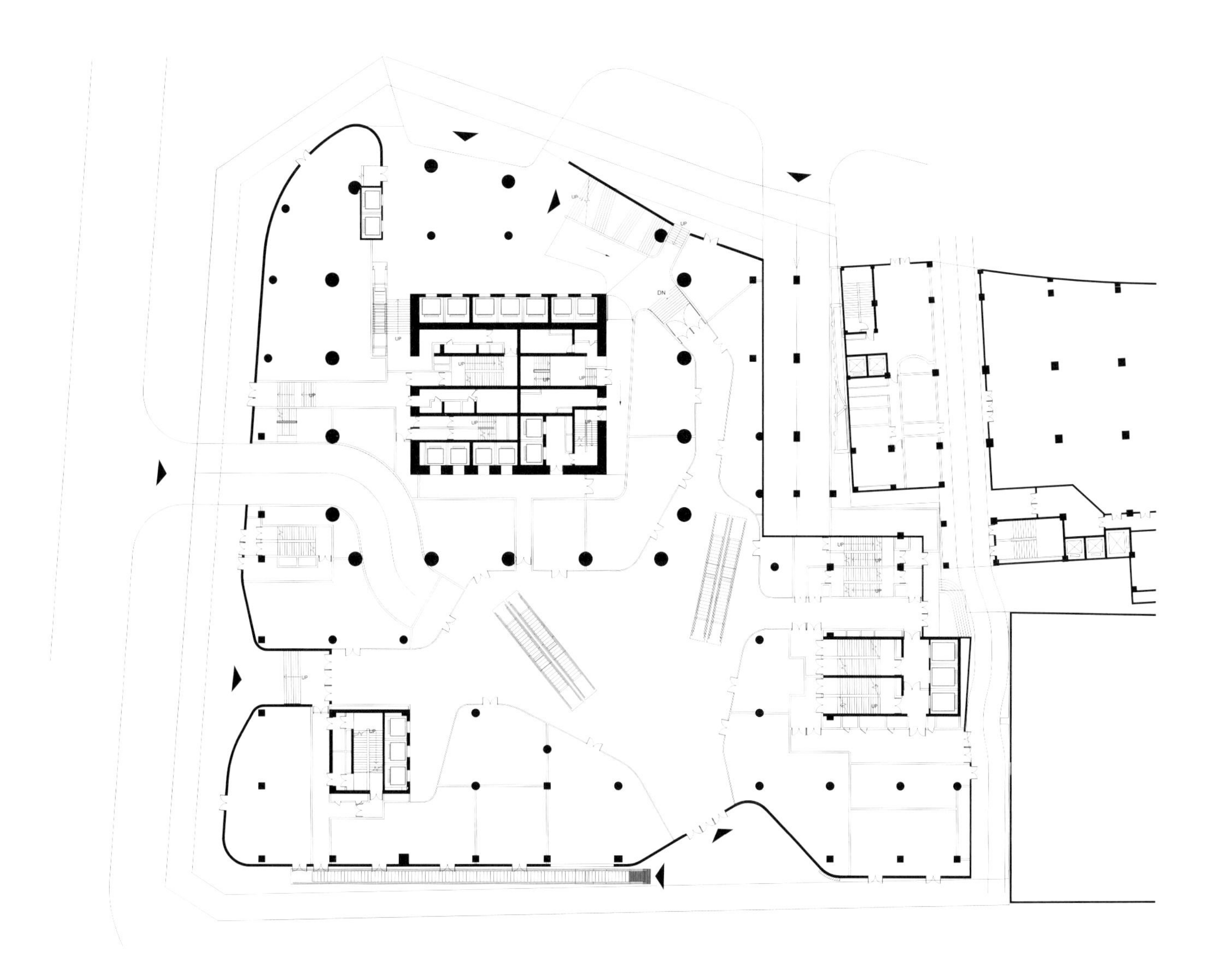

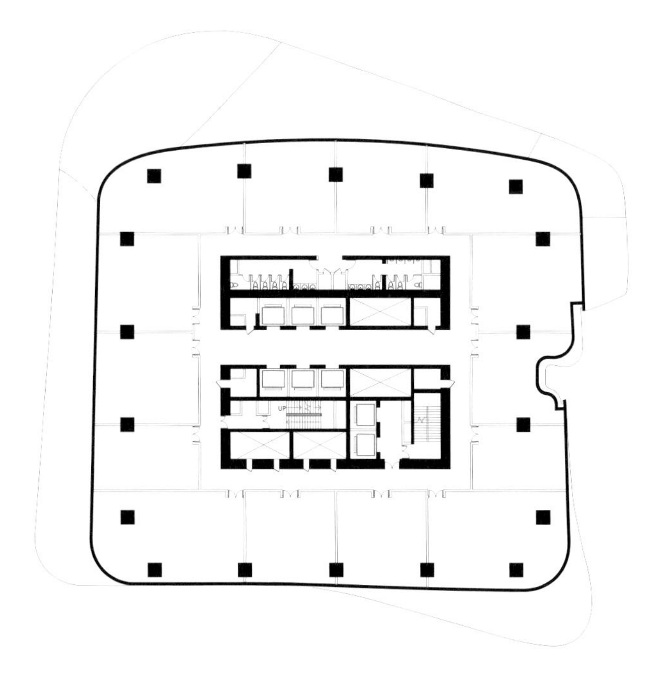

10
8 9
11

8. 建筑外观人视效果图
External view
9. 立面图
Elevation
10. 一层平面图
Level 1 floor plan
11. 办公低区标准层平面图
Typical office floor plan

武汉恒隆广场

Heartland 66, Wuhan

武汉，历来被称为“九省通衢”之地，它是中国中部地区的中心城市，得“中”独厚。作为一个“老牌”城市，如今的武汉也正朝着国际都市的方向迈进。武汉恒隆广场地处武汉市繁华的商贸枢纽硚口区京汉大道，邻近轻轨及地铁网络。通过精心打造的从地面缓缓升起的建筑体量与充满活力的入口，项目不仅与周边环境和谐相融，也拥有自己独特的个性。

武汉恒隆广场的设计概念“无限循环”源于中国结，通过购物中心的人流动线和屋顶花园的造型得以诠释。商场内的蜿蜒动线于水平和垂直方向把主力店和活动空间无缝串联起来。设计通过从建筑物内贯穿循环到屋顶花园外部的流线，生动形象地反映出建筑空间内外的连通性，让室内与室外一起共同沐浴着阳光。办公楼建筑主体由底部的多个体块巧妙地交织旋转直到塔冠，同样也体现了“无限循环”的理念。

如宝石般的建筑造型传递着优雅、宏伟的美学语言。通过对武汉的社会、环境和文化方面的深刻理解，Aedas 设计的武汉恒隆广场，不仅传承了历史遗产，同时也延续着武汉的城市个性，并切合武汉市民的需求。

1. 武汉的现代城市风光
The modern city view of Wuhan
2. 整体效果图
External view
3. 鸟瞰效果图
Aerial view

项目信息 | PROJECT INFORMATION

地　　点：中国武汉	Location: Wuhan, PRC
建筑面积：432 500m^2	Gross floor area: 432,500 m^2
设计时间：2013—2017 年	Design time: 2013-2017
建成时间：2020 年	Completion year: 2020
项目功能：综合体	Sector: Mixed-use
业　　主：恒隆地产有限公司	Client: Hang Lung Properties Ltd.
主要设计人：林静衡，祈礼庭	Directors: Christine LAM, David CLAYTON

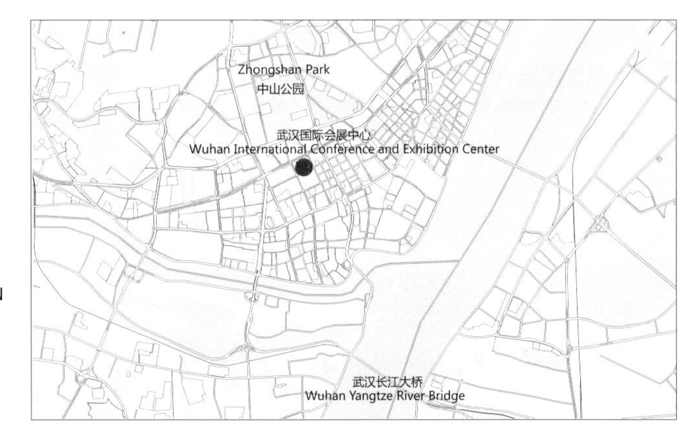

Wuhan, known as the “thoroughfare of nine provinces,” is the most important city in central China. Known as an ancient city, Wuhan nowadays is stepping toward the future as a cosmopolis. Heartland 66 is located on Jinghan Avenue in the bustling Qiaokou District, the commercial and business heart of Wuhan, with light rail and mass transit railways running through the area. By increasing permeability through buildings that gently rise from the ground, with dynamic components inserted at the entrance, the design opens up the building into the public realm while maintaining its individuality.

The design of Heartland 66 borrows the concept of the infinity loop, as well as the Chinese art of the knot tie. The infinite form of the knot is represented by the circulation and roof garden, connecting the anchors and activity spaces with seamlessly flowing arcades. The landscape design reinforces the arrival experience at the ground floor entrances and sunken plaza areas, and it also strongly reflects the connectivity between inside and outside as the route circulating up through the building and terminating at the external roof garden. The infinity loop concept was further adopted and adapted to the specific requirements of an office tower, with the resulting form of multiple masses subtly entwining and shifting at the base before sweeping and stepping into the crown of the tower.

The gem-like aesthetics of the building convey a sense of grace and grandeur. Heartland 66 was designed with deep understanding of Wuhan’s social, environmental and cultural aspects, focusing not just on its heritage, but also embracing the needs of Wuhan citizens while maintaining the identity of the city.

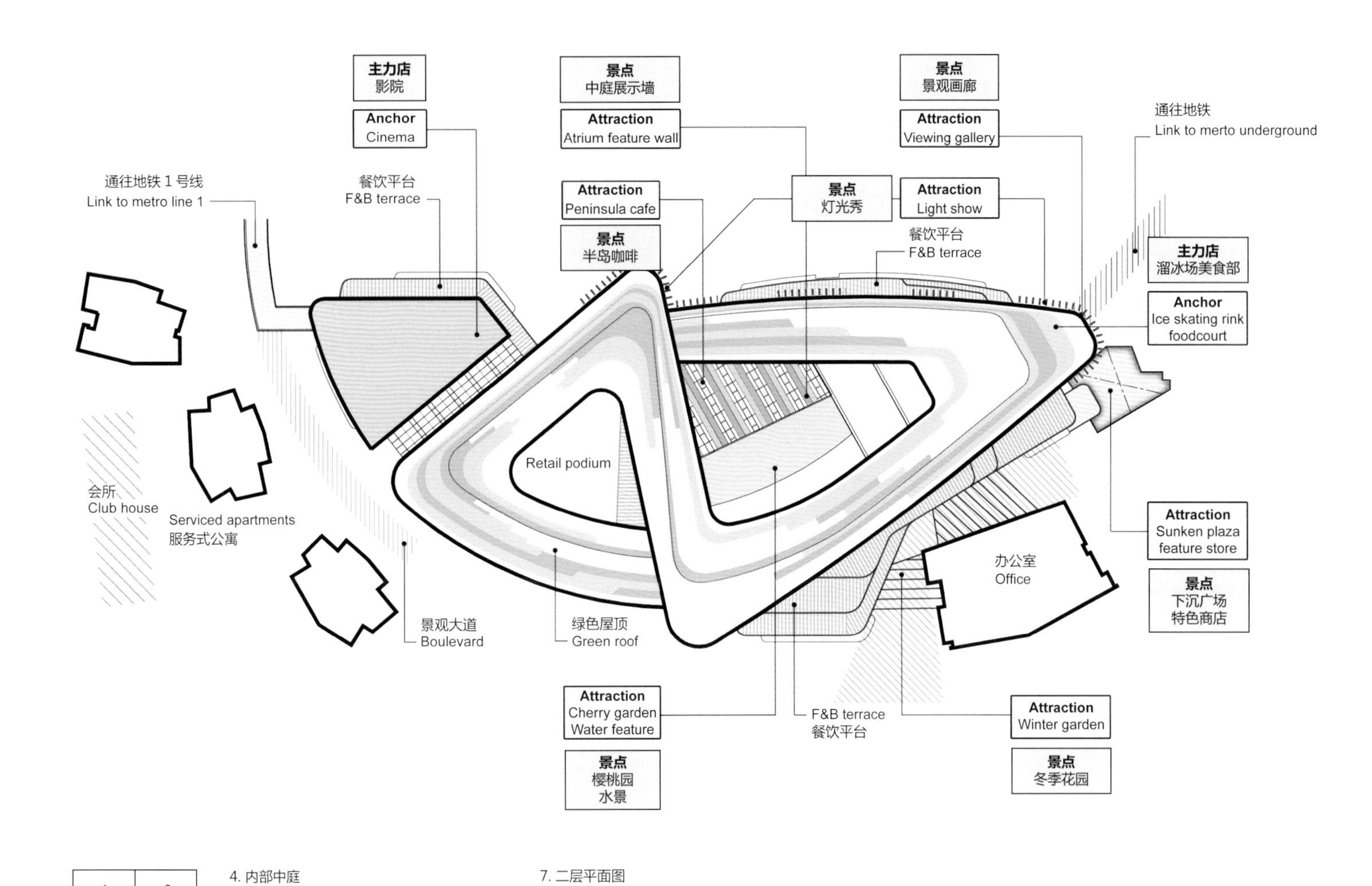

4 5	6 7

4. 内部中庭
Atrium
5. 项目区位图
Location map
6. 分析图
Analysis diagram

7. 二层平面图
Level 2 floor plan

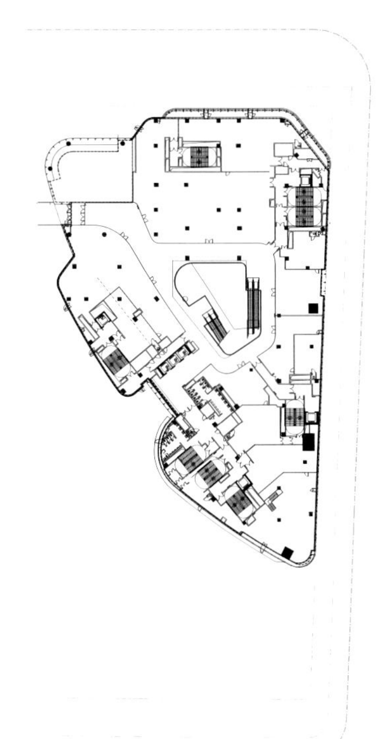

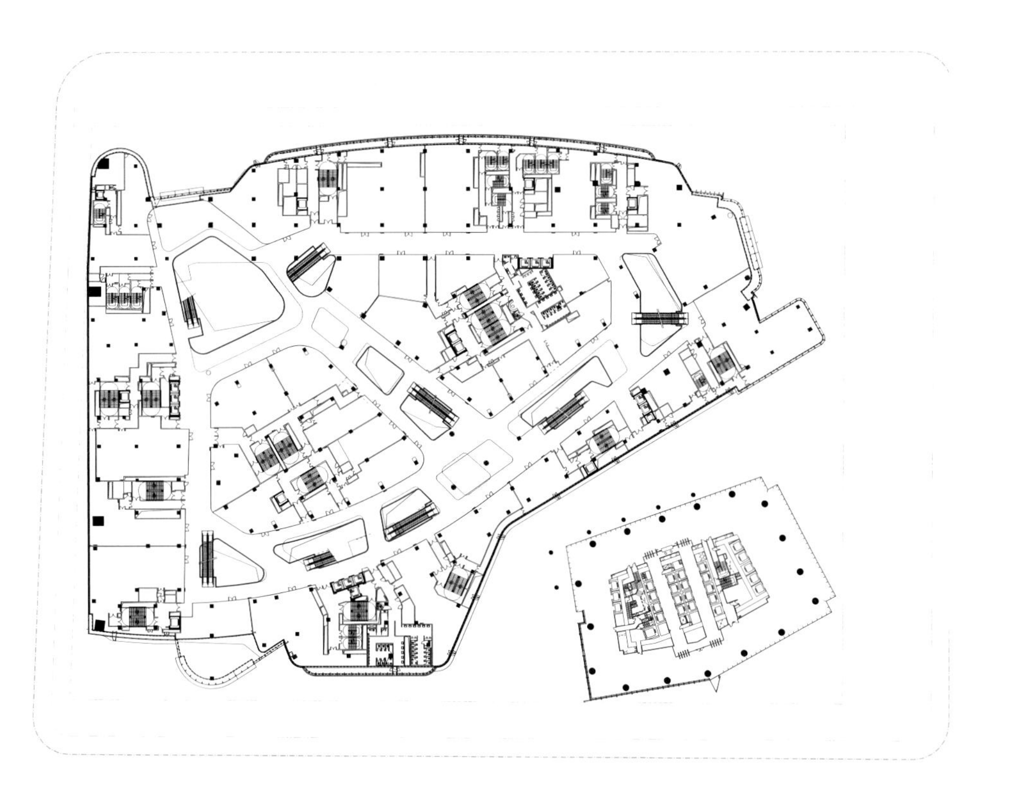

中国大陆长江以南
SOUTH OF YANGTZE RIVER

中国大陆长江以北 NORTH OF YANGTZE RIVER

- 2 3 4 7 大连恒隆广场 Olympia 66, Dalian
- 2 4 6 北京大望京综合开发项目 Da Wang Jing Mixed-use Development, Beijing
- 3 4 北京大兴 3 及 4 地块项目 Daxing Plots 3 and 4, Beijing
- 1 3 6 北京新浪总部大楼 Sina Plaza, Beijing
- 1 2 3 4 6 北京北苑北辰综合体 North Star Mixed-use Development, Beijing
- 2 3 4 7 青岛金茂湾购物中心 Jinmao Harbour Shopping Center, Qingdao
- 1 2 4 成都恒大广场 Evergrande Plaza, Chengdu
- 1 7 重庆新华书店集团公司解放碑时尚文化城 Xinhua Bookstore Group Jiefangbei Book City Mixed-use Project, Chongqing
- 2 3 4 7 武汉恒隆广场 Heartland 66, Wuhan

中国大陆长江以南 SOUTH OF YANGTZE RIVER

- 3 6 7 上海星荟中心 Shanghai Landmark Center, Shanghai
- 1 2 3 4 6 上海龙湖虹桥项目 Longfor Hongqiao Mixed-use Project, Shanghai
- 6 7 上海虹桥世界中心 Hongqiao World Center, Shanghai
- 2 3 4 6 7 无锡恒隆广场 Center 66, Wuxi
- 3 7 苏州西交利物浦大学中心楼 Xi'an Jiaotong-Liverpool University Central Building, Suzhou
- 2 3 义乌之心 The Heart of Yiwu, Yiwu
- 2 3 6 广州南丰商业、酒店及展览综合大楼 Nanfung Commercial, Hospitality and Exhibition Complex, Guangzhou
- 3 8 广州邦华环球贸易中心 Bravo PARK PLACE, Guangzhou
- 3 珠海粤澳合作中医药科技产业园总部大楼 Headquarters, Traditional Chinese Medicine Science and Technology Industrial Park of Co-operation between Guangdong and Macao, Zhuhai
- 3 8 深圳宝安国际机场卫星厅 Shenzhen Airport Satellite Concourse, Shenzhen
- 1 7 珠海横琴国际金融中心 Hengqin International Financial Center, Zhuhai
- 2 3 4 珠海横琴中冶总部大厦（二期） Hengqin MCC Headquarters Complex (Phase II), Zhuhai

中国改革开放从南方起步，从沿海开始，经过 40 年的发展，无论是城市化水平还是经济体量都达到一个新的高度，城市建设中留下了中国的经验和独特的中国模式，无论是商业综合体、金融中心、学校、文化建筑，都以一种新的思考与姿态展现在南方城市的活跃地区。Aedas 在这些地区的作品无疑成为了城市活力区域的标志。

China's reform and opening up began in the southern coast. After 40 years of development, both level of urbanisation and size of the economy have reached a new height. Urban development led to China's experience and unique development model. Mixed-use complexes, financial centers, school and cultural facilities are presented in a new way of thinking and form in southern cities. Aedas' projects in this region have undoubtedly become urban icons.

中国香港地区和中国台湾地区 HONG KONG AND TAIWAN REGIONS OF CHINA

- ❶❺❽ 香港西九龙站 Hong Kong West Kowloon Station, Hong Kong
- ❷❸❺❻ 港珠澳大桥香港口岸旅检大楼 Hong Kong-Zhuhai-Macao Bridge Hong Kong Port - Passenger Clearance Building, Hong Kong
- ❶❺❽ 香港国际机场中场客运廊 Hong Kong International Airport Midfield Concourse, Hong Kong
- ❶❺❽ 香港国际机场北卫星客运廊 Hong Kong International Airport North Satellite Concourse, Hong Kong
- ❹ 香港富临阁 The Forum, Hong Kong
- ❸❼ 台北砳建筑 Lè Architecture, Taipei
- ❸ 台中商业银行企业总部综合项目 Commercial Bank Headquarters Mixed-use Project, Taichung

一带一路 BELT AND ROAD

- ❸❹❽ 新加坡星宇项目 The Star, Singapore
- ❸❽ 新加坡 Sandcrawler Sandcrawler, Singapore
- ❽ 阿联酋迪拜 Ocean Heights Ocean Heights, Dubai, UAE
- ❺❽ 阿联酋迪拜地铁站 Dubai Metro, Dubai, UAE
- ❽❾ 英国唐卡斯特 Cast 剧院 Cast, Doncaster, UK

九项设计理念 NINE POINTS OF DESIGN IDEAS

1. 《国家新型城镇化规划（2014—2020 年）》 NATIONAL NEW URBANISATION PLAN (2014-2020)
2. 高密度城市枢纽 HIGH DENSITY CITY HUBS
3. 具有通透性和连接性的设计 POROUS AND CONNECTED DESIGNS
4. 新型商业零售 THE NEW RETAIL
5. 基础设施设计 INFRASTRUCTURE DESIGN
6. 中国的特大城市与城市设计 CHINA MEGAPOLIS AND URBAN DESIGN
7. 文化和故事 CULTURE AND STORY
8. 一带一路 BELT AND ROAD
9. 城市改造与修复 URBAN RENEWAL AND RESTORATION

上海星荟中心

Shanghai Landmark Center, Shanghai

上海苏州河是黄浦江的重要支流之一，苏州河沿岸以其独有的历史文化和滨水生活，形成了充满活力的商业、居住环境。上海星荟中心位于苏州河北岸，东向及南向坐拥北外滩与黄浦江景色，享有极佳的自然与城市景观。而项目的北侧主要是低中层居住区，其中包含众多历史保护建筑。在这一新建综合发展项目中，建筑将以其灵活的使用方式，形成滨水商业中心与居住区的完美衔接。

项目的设计灵感来自中国传统建筑中的“格栅”意象。这种古老的东方元素用当代语言进行全新诠释后呈现出令人耳目一新的效果，同时还强化了与周边民宅和历史建筑的联系，使项目完美融入区域建筑风格之中。

与典型的玻璃塔楼相比，该建筑的塔楼外立面降低了光的反射率，这是为了减少对周边环境的光污染。方正的建筑外形既奠定出了大气开阔的整体基调，又继承了周边历史文化建筑的风格。“格栅”这一鲜明的东方元素从建筑外立面延伸到内层空间，灯光通过栅格，给建筑内部带来了丰富的光影变化。建筑的竣工重新定义了该地区的城市天际线，它成为了苏州河畔上海天际线中的地标建筑。

1. 上海浦东天际线
Skyline of Pudong, Shanghai
2. 从苏州河方向看上海星荟中心
View from Suzhou Creek
3. 上海星荟中心建筑外观
External view

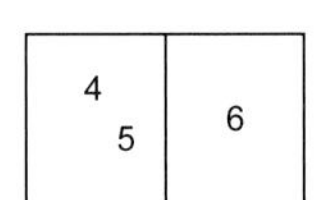

4. 手绘图
Sketch
5. 项目区位图
Location Map
6. 外立面细部
Facade details

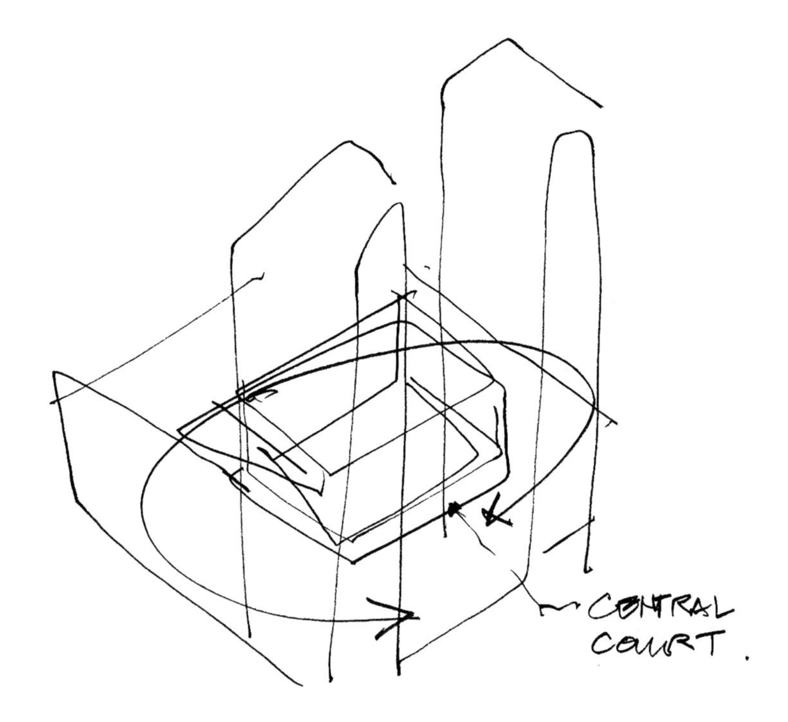

项目信息 | PROJECT INFORMATION

地　　点：中国上海
建筑面积：161 000m²
建成时间：2017 年
项目功能：综合体
业　　主：香港建设（控股）有限公司、上海广田房地产开发有限公司
主要设计人：柳景康

Location: Shanghai, China
Gross floor area: 161,000 m²
Completion year: 2017
Sector: Mixed-use
Client: HKC (Holdings) Ltd., Shanghai Guangtian Real Estate Development Company Limited
Director: Cary LAU

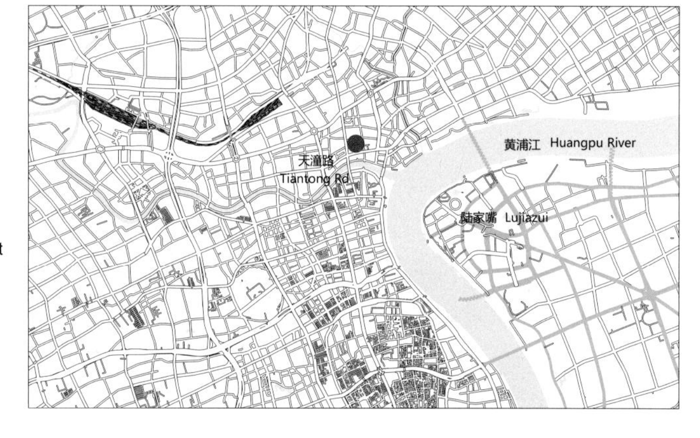

Suzhou Creek in Shanghai is one of the most important tributaries of the Huangpu River. The area along Suzhou Creek has unique history, culture and waterfront life, including a dynamic commercial and residential environment. Shanghai Landmark Center is located on the north bank of Suzhou Creek with excellent views towards the northern Bund and the Huangpu River. The site is surrounded by low-rise residential blocks to the north, including a number of historic buildings. The new complex creates a synergy for the commercial hub and the neighbourhood, connecting the old and new sides of Shanghai.

Inspired by traditional windows in nearby historic buildings, the facade interprets this typical Chinese element in a contemporary form, creating an impressive motif. This allows it to forge a closer relationship with the surrounding housing and historic buildings, blending itself into the community.

This modern interpretation of historic windows on the tower facade allows less reflectance compared to typical glass towers, and relieves its neighbouring buildings from light pollution and disturbance. The square-shaped design not only underlines the generous character of the project, it also inherits the style of the surrounding historic buildings. The signature motif extends to the retail blocks, serving as an ornamental screen on the facade while also adding interesting, animated light patterns to the inner layer of the building. The project is no doubt a new landmark on the bank of Suzhou Creek.

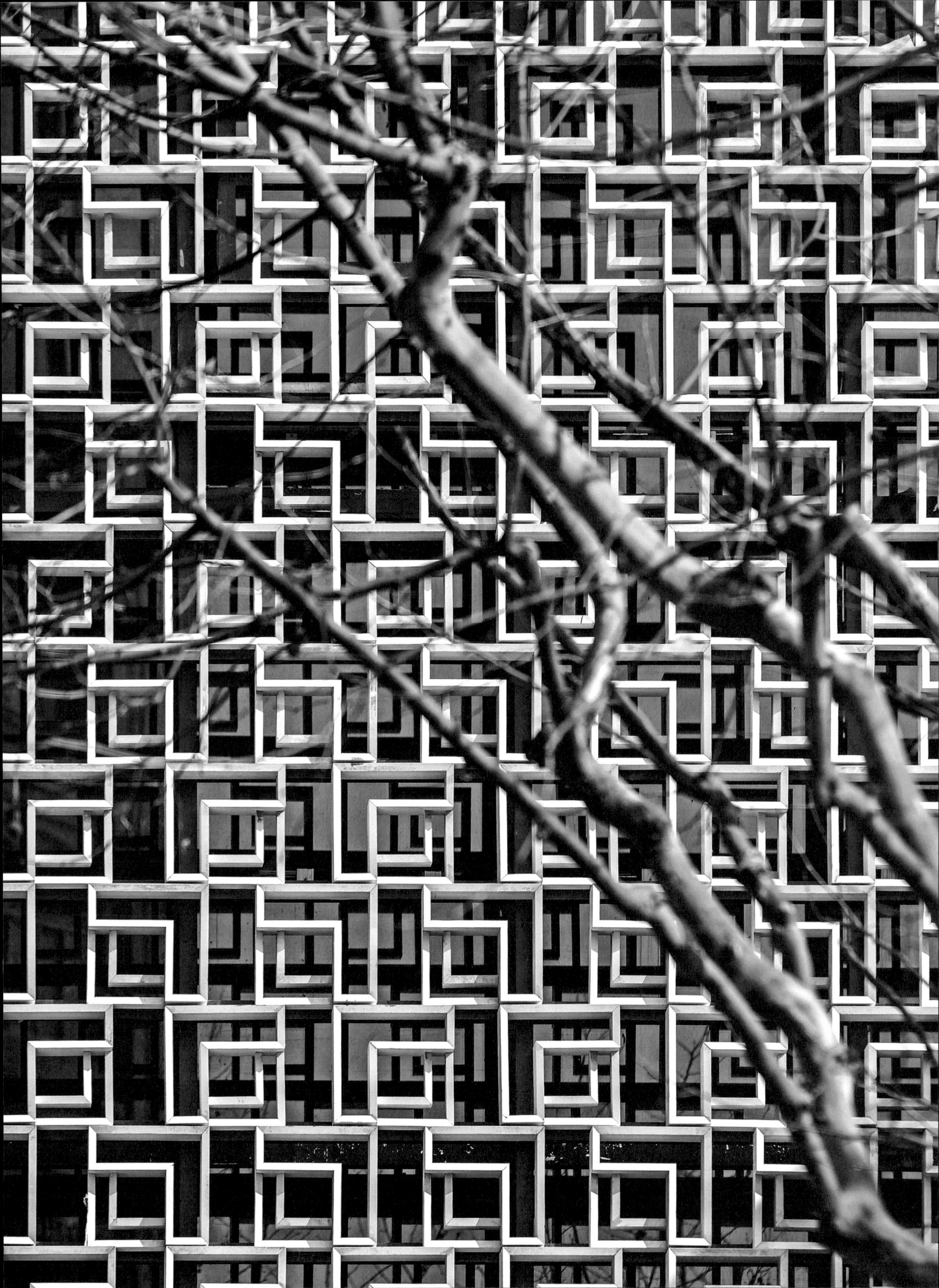

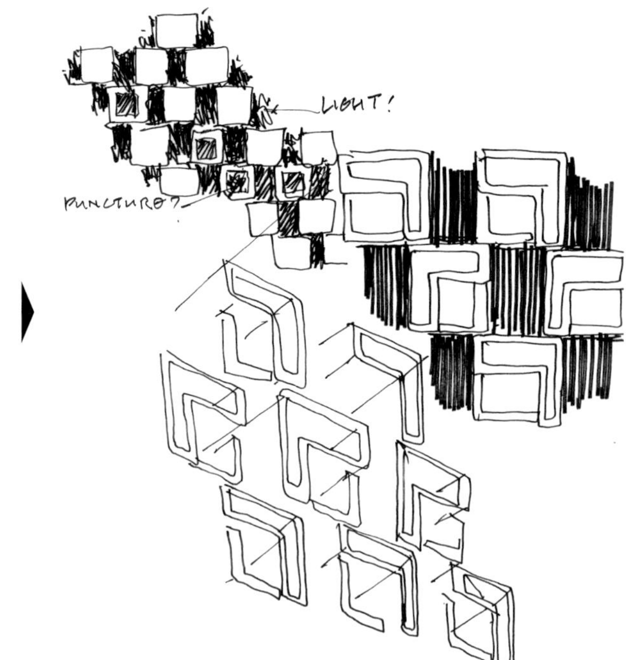

7	9
8	10

7. 将“格栅”意象融入建筑外立面设计
Traditional Chinese windows inspired the facade desgin
8. 立面图
Elevation
9. 星荟中心建筑入口
Entrance
10. 一层平面图
Level 1 floor plan

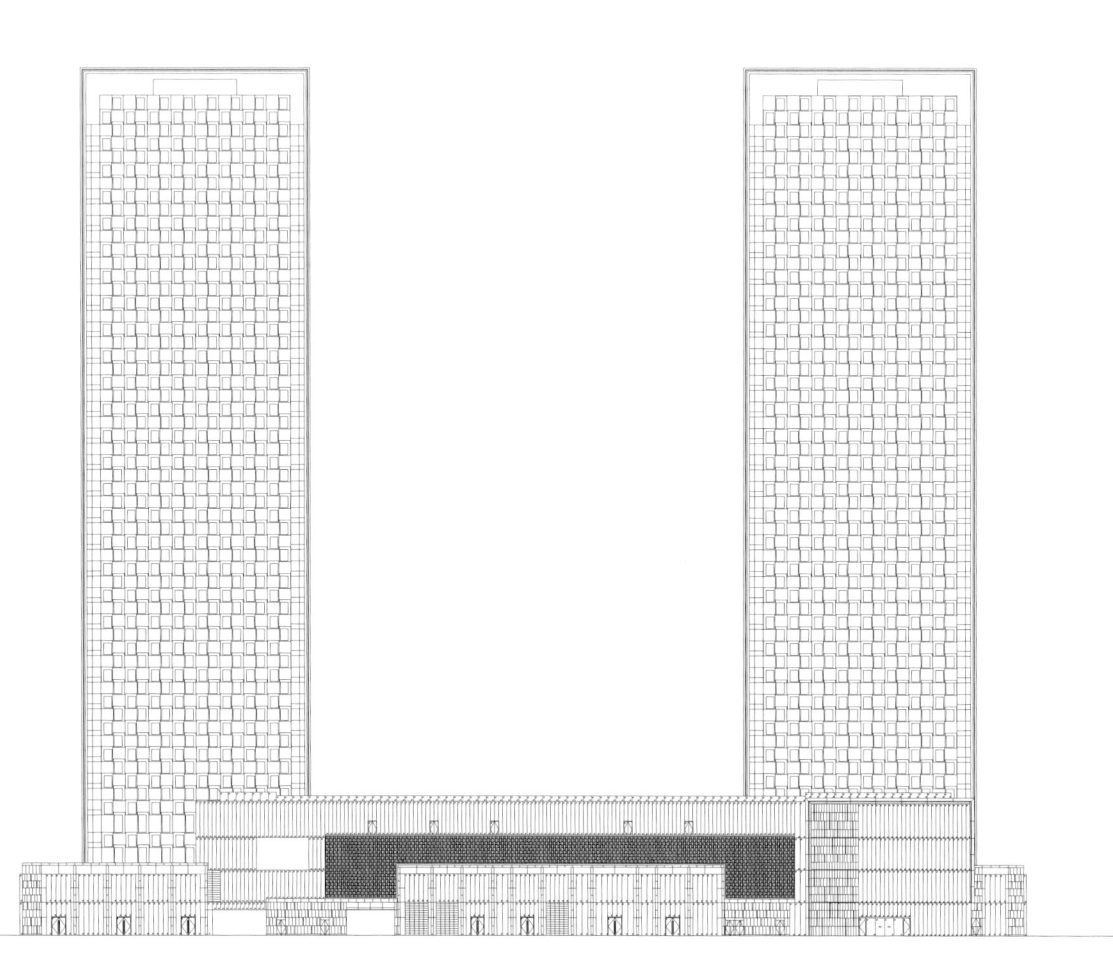

LANDMARK Center

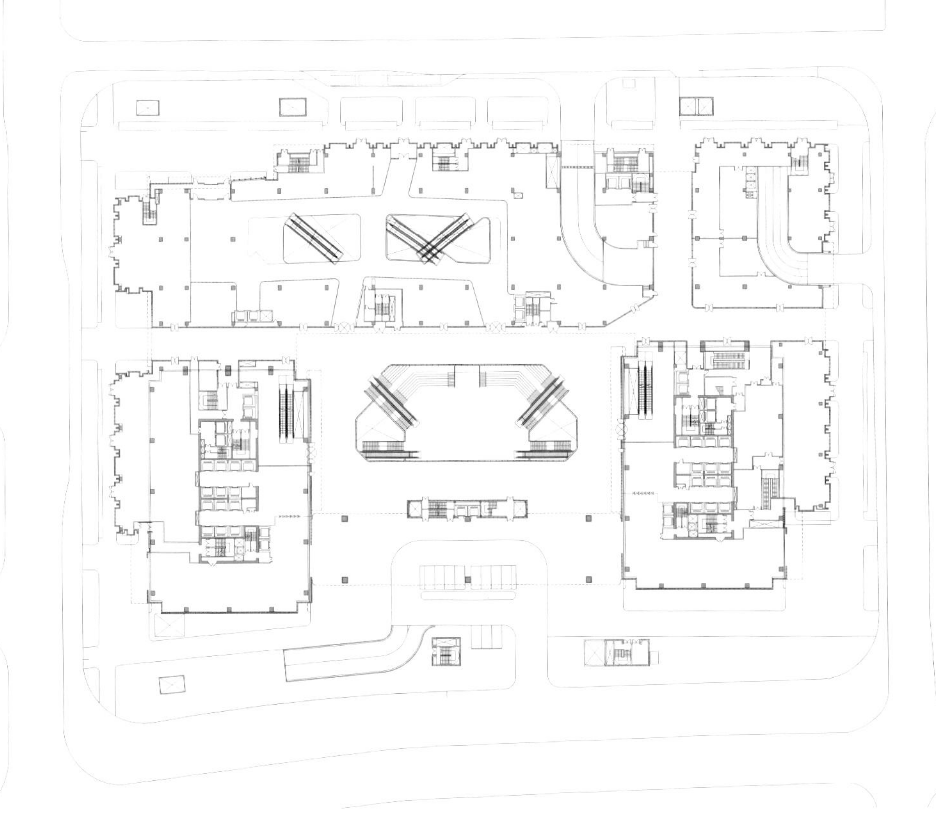

上海龙湖虹桥项目

Longfor Hongqiao Mixed-use Project, Shanghai

龙湖虹桥天街位于上海西部的大虹桥商务区，毗邻虹桥机场。该区域作为 2010 年上海世博会主要基础设施建设的一部分，包含有虹桥机场和虹桥火车站，被规划为上海这个未来世界级城市群龙头的核心枢纽，而其势如破竹的发展速度以及商务区的雏形初具，都见证着这一交通枢纽转化为集聚商业物流、信息流与资金流的世界级宜居之地。

项目包含商业零售、精品商业、SOHO 办公和酒店等功能。作为设计重点的东侧商业区，灵感来自于上海里弄和中国元素“盘龙”。江南传统的吴越文化与西方传入的工业文化相融合，形成了上海特有的“海派文化”，传统的里弄“石库门”建筑正是这一文化的典型空间形态，由里弄与庭院组成的空间尺度宜人、归属感强。设计采取“里弄”+“庭院”的空间模式，将线性的商业街与购物中心融为一体，打破综合商业体的大体量，结合方向性明确的购物流线，增加人流吸引力，提升商业效益。同时，设计配合建筑形态形成一系列各具特色的户外空间，打造出多样性的公共场所，并使商业街立面充满活力。此外，空中步廊是该项目的另一特点，二层步廊系统将该项目与虹桥枢纽紧密联系，并贯联起各楼群及周边项目，力促虹桥核心区实现“跨街区整体化、立体化”。

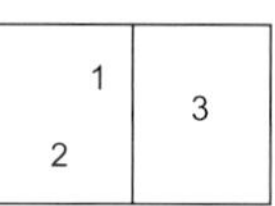

1. 上海虹桥商业区的天际线
The skyline of Hongqiao commercial district in Shanghai
2. 鸟瞰龙湖虹桥天街
Aerial view
3. 独具特色的空中步廊系统使商业与办公楼互相连接，将直通化、商务化、景观化、社区化、标志化等多个目标整合为一体
Different programmes are well connected with bridges and walkways, creating a holistic development

虹桥天街租赁中心

项目信息 | PROJECT INFORMATION

地　　点：中国上海
建筑面积：513 800m^2
设计时间：2012—2013 年
建成时间：2016 年
项目功能：综合体
业　　主：上海龙湖置业发展有限公司
主要设计人：林静衡，祈礼庭

Location: Shanghai, PRC
Gross floor area: 513,800 m^2
Design time: 2012-2013
Completion year: 2016
Sector: Mixed-use
Client: China Longfor Co., Ltd.
Directors: Christine LAM, David CLAYTON

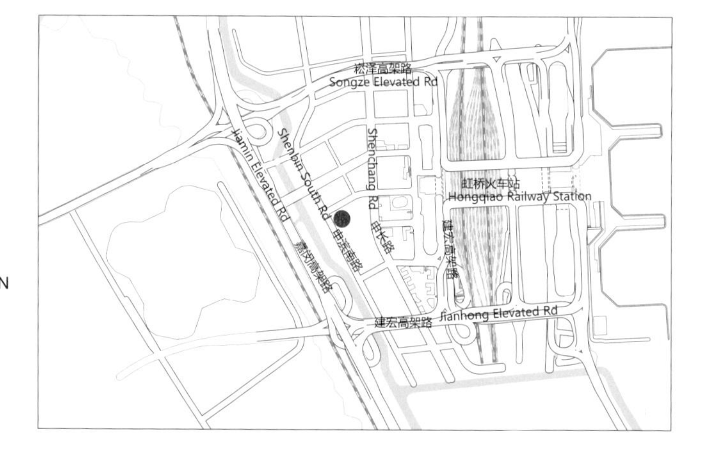

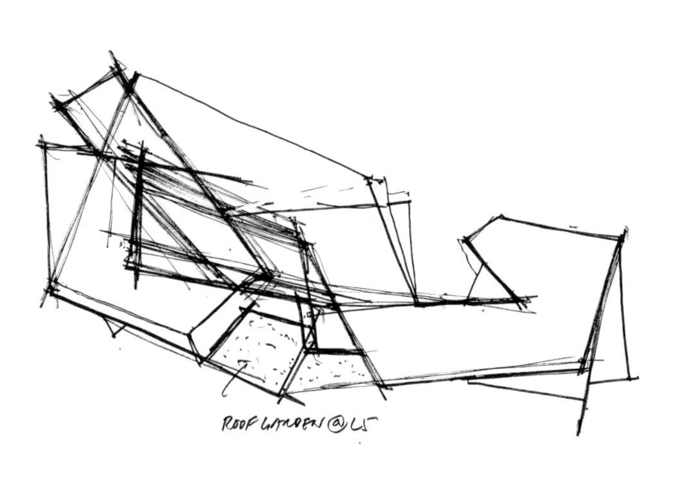

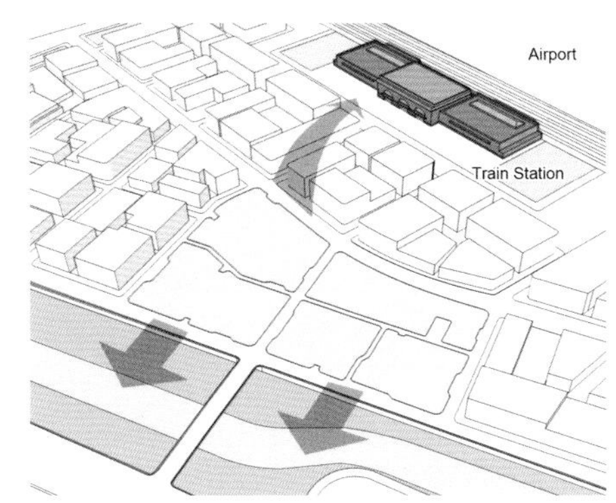

跟基地文脉的关系：
基地作为中心连接滨水步道，火车站，机场

Connections to the water promenade, train station and airport

高街零售店面呼应毗邻的位于主动线上的建筑立面

High-street retail frontage along the main road

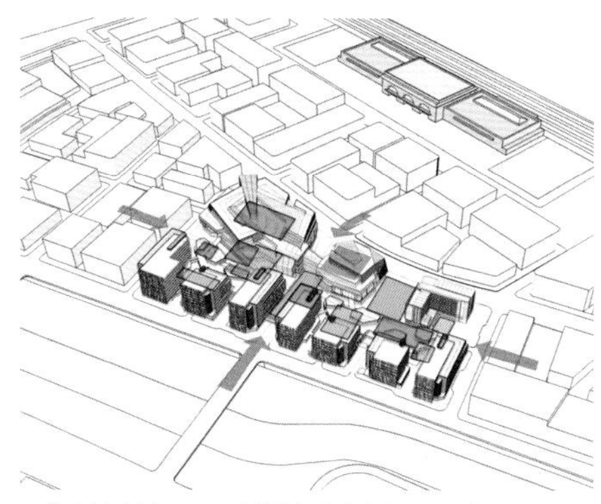

口袋型空间被嵌入不同功能之间的空隙中，并作为节点跟流线结合在一起

Pocket spaces to enhance circulation

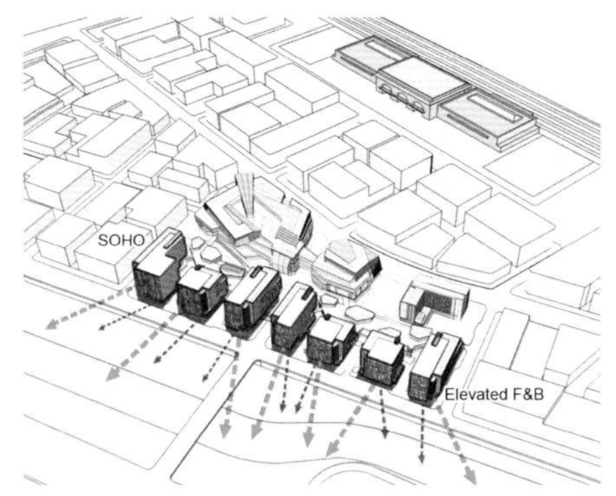

裙房架高以确保 SOHO 朝向滨水公园的最佳视线

Elevated podium ensures best view for SOHO

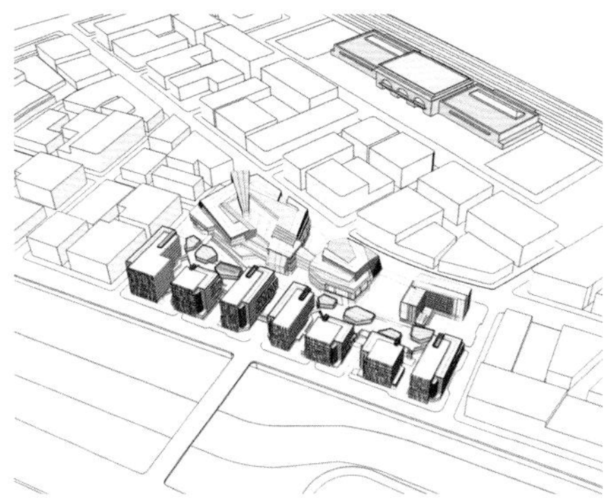

错落的建筑体量打造适合行人尺度的社区空间

Staggered building masses create human-scale community spaces

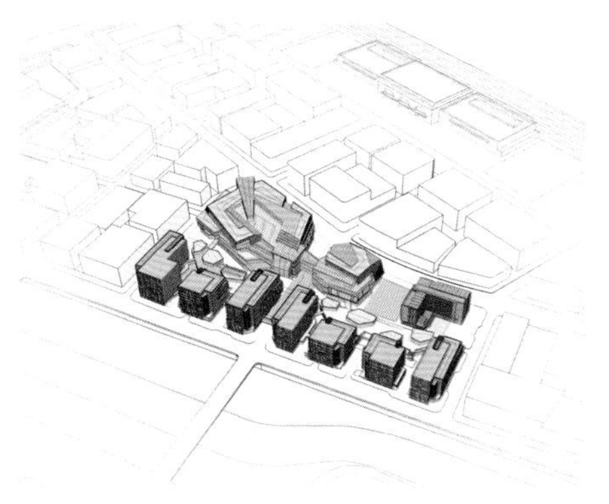

设计实现了不同业态的和谐共存

The design allows all functional programmes to co-exist in harmony

Longfor Hongqiao Mixed-use Project is located in the Greater Hongqiao Business District, in western Shanghai, next to the airport. This area is one of the main hubs of development built for the 2010 Shanghai World Expo, and it includes Hongqiao International Airport and Hongqiao Railway Station. The area is planned to become a core hub for Shanghai, which is looking to be the sixth largest city in the world. With an ambitious appetite for development, this quickly growing business district is becoming a transportation hub and a world-class livable city for business logistics, information flow and capital flow.

The project includes retail shops, SOHO office facilities and a hotel. The design focuses on the commercial district on the east side of the development, and is inspired by the alleyways of Shanghai, as well as the “coiled dragon” that is often found in Chinese mythology. The harmony of Jiangnan’s traditional Wuyue culture and Western-influenced industrial culture formed a unique Shanghainese culture. This culture is represented by the architecture of housing known as Shikumen, found in the city’s alleyways. Built around alleyways and courtyards, this type of housing features a comfortable scale and a strong sense of community. This project’s design made use of the spatial patterns of alleyways and courtyards, merging linear shopping streets into the shopping center as a whole. Compared to the huge mass of the integrated commercial complexes that are common in Shanghai, this design has a gentler scale and clearer circulation, so as to attract consumers and improve commercial efficiency. Meanwhile, the design has created unique and diverse public outdoor spaces that bring life to the shopping area. Another feature is the sky bridge on the second floor that is tightly connected to other developments in the Hongqiao hub, further encouraging the development of the Hongqiao core area, with three-dimensional circulation between the blocks.

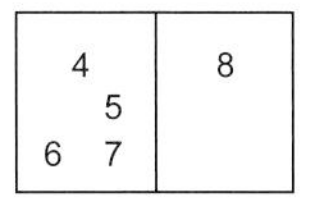

4. 酒店建筑外观
Exterior view of Hotel Indigo Shanghai Hongqiao
5. 项目区位图
Location map
6.7. 手绘图
Sketches
8. 龙湖虹桥天街分析图
Analysis diagrams

天街
TOP SPORTS
天街

9	11
10	12

9. 建筑夜景，白色铝板配合象征龙鳞的 LED 灯，动态的外形点出商场潮流动感的定位
Night view of the shopping center
10. 内部中庭是充满活力的购物及休闲空间
Atrium of the shopping center
11. 一层平面图
Level 1 floor plan
12. 东立面图
East elevation

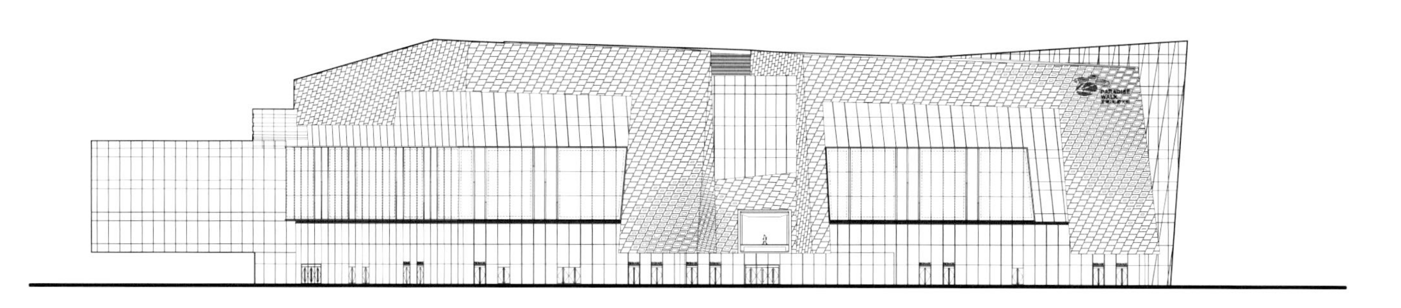

上海虹桥世界中心

Hongqiao World Center, Shanghai

有着“东方巴黎”美誉的上海是中国的中心城市。作为大虹桥板块的西门户，虹桥世界中心为全球最大的国家会展中心和虹桥中央商业区提供顶级的配套设施，倾心打造城市名片——“向阳花”。整个规划以呼应国展中心“四叶草”造型为出发点，神从形似，各个建筑单体以“花心”“花瓣”“绿叶”的形式有机排列组合，组成“向阳花”的形态，与“四叶草”轴心对位，构成区域内最具辨识性的建筑群。

虹桥世界中心与国家会展中心合为一体，由 Aedas 设计的项目内容包括五星级酒店、购物中心以及展示中心。酒店和购物中心成为整个项目的入口，展示中心作为最靠近会展中心出口的建筑，由人行天桥与会展中心相连接。展示中心设计受“叶子”启发，形成一个相当柔软和动态的外观。所有的功能空间都围绕着中庭展开，时而宽敞，时而狭窄，天庭引入了自然采光，产生出不同的曲面和光影，形成独特的仿生的立面效果。

1	3
2	4

1. 上海城市景观
Cityscape of Shanghai
2. 俯瞰虹桥世界中心及国家会展中心
Aerial view
3. 酒店外观
External view of Primus Hotel Shanghai Hongqiao
4. 项目区位图
Location map

项目信息 | PROJECT INFORMATION

地　　点：中国上海	Location: Shanghai, PRC
建筑面积：172 000m^2	Gross floor area: 172,000 m^2
设计时间：2014 年	Design time: 2014
建成时间：2018 年	Completion year: 2018
项目功能：综合体	Sector: Mixed-use
业　　主：绿地集团	Client: Greenland Group
主要设计人：纪达夫	Director: Keith GRIFFITHS

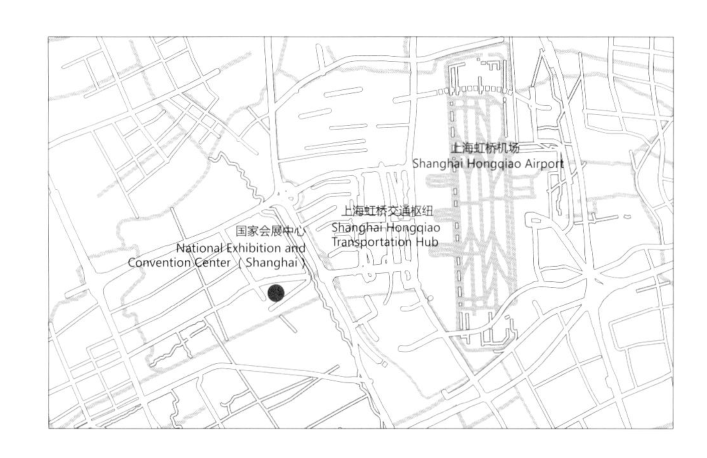

Shanghai, the so-called "Paris of the Orient," is one of the most important cities in China. As the western gate of Great Hongqiao Area, the Hongqiao World Center offers high-end supporting facilities to the National Exhibition and Convention Center and the Hongqiao Central Business District. In response to the clover leaf concept of the National Exhibition and Convention Center, the leaf-shaped buildings of the Hongqiao World Center are organised in a way that promotes smooth circulation.

The Hongqiao World Center comprises a five-star hotel, a shopping center and a sales gallery. The hotel and shopping center will work together to form a very welcoming entrance to the whole development. The sales gallery, as the first completed building, is connected to the National Exhibition and Convention Center by a pedestrian bridge. The gallery, shaped as a leaf, has a soft and dynamic appearance. All the functional spaces are organised around a central exhibition atrium space. The atrium consists of spaces contrasting in size, with a skylight introducing the natural light which creates different tones and depth of shadows.

5 6 7 8	9

5. 设计灵感及手绘图
Design concept and sketch
6. 商场与酒店空间流线设计分析
Circulation analysis
7. 酒店首层平面图
Level 1 floor plan of hotel
8. 商场首层平面图
Level 1 floor plan of shopping center
9. 展示中心外观
External view of the sales gallery

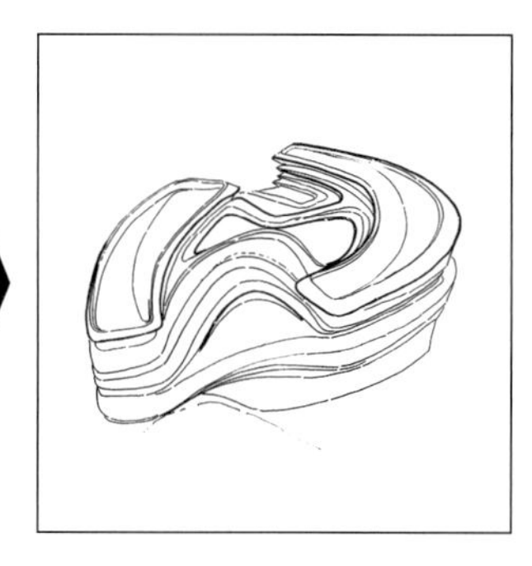

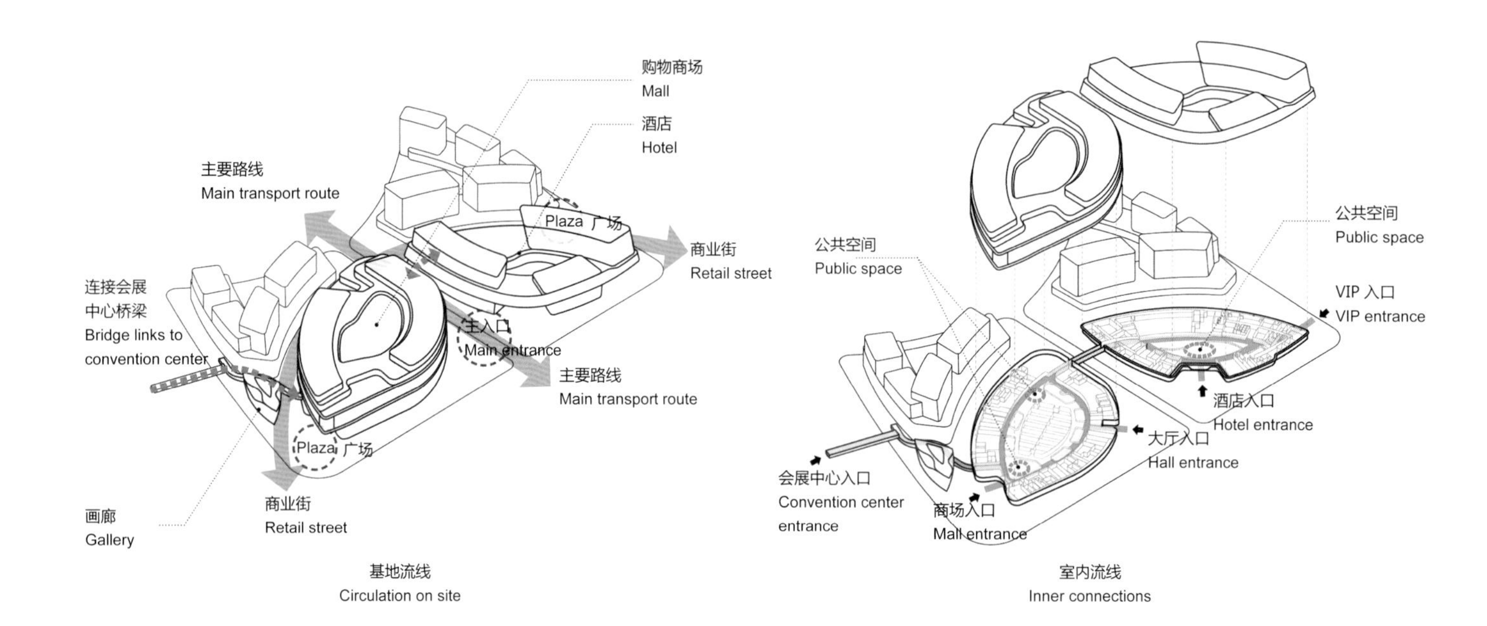

基地流线
Circulation on site

室内流线
Inner connections

无锡恒隆广场

Center 66, Wuxi

建筑在界定自身空间的同时，亦需精心实现城市新旧空间的转译。无锡恒隆广场项目所处城隍庙历史街区有着大量的历史建筑群，建筑师将这些历史建筑保留下来，成为由大体量的写字楼与商场围合的公共广场的文化内核。恒隆广场地处无锡两条最繁华的商业街——中山路和人民中路交界口，商场在地库一层设有隧道，直接连通地铁站，提供方便快捷的交通接驳。在此，不同时代的建筑展开穿越时光的对话，空间上形成一条有趣的时间轴。

传统与现代的和谐是整个项目的出发点。设计师透过“动态”手法创造出历史建筑物和全新城市建筑的对话空间。设计理念取自中国书法的精髓，以饱满流畅的曲线增强了项目的形体和空间感。办公楼造型为两个狭窄的盒子并肩错位相连，主立面的幕墙从底部至顶部对角沿阶而上。6 层高的商业裙楼用眩目而复杂的玻璃立面代替普通幕墙框架，形成光滑的全玻璃表皮。规划设计将各功能活动空间紧密联系起来，其中的写字楼犹如“门户”串联并延续动线至室外广场。整个项目以时代精神重塑城市空间，也珍藏起城市的历史。

1 2 3

1. 无锡城市景观
Cityscape of Wuxi
2. 建筑外观
External view
3. 俯瞰商业裙楼
Aerial view

项目信息 | PROJECT INFORMATION

地　　点：中国无锡
建筑面积：376 800m^2
设计时间：2006—2010 年
建成时间：2014 年
项目功能：综合体
业　　主：恒隆地产有限公司
主要设计人：林静衡，祈礼庭

Location: Wuxi, PRC
Gross floor area: 376,800 m^2
Design time: 2006-2010
Completion year: 2014
Sector: Mixed-use
Client: Hang Lung Properties Ltd.
Directors: Christine LAM, David CLAYTON

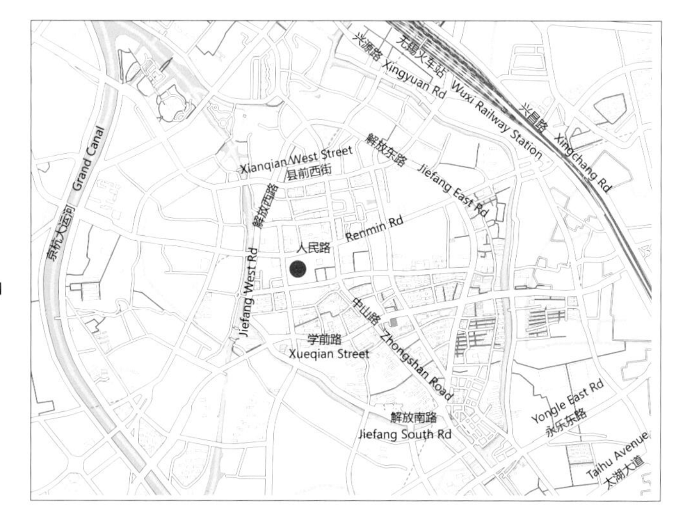

An architecture defines its own space and is often an interpreter of old and new urban spaces. Center 66 is located in the historic district around the ancient Chenghuang Temple complex. Historic buildings on site are preserved to become a cultural core for the public plaza, which is enclosed by large-scale office buildings and shopping mall. Located at the junction of the two busiest shopping streets in Wuxi, Zhongshan Road and Renminzhong Road, the basement floor of the complex includes a tunnel connecting to the metro station, providing

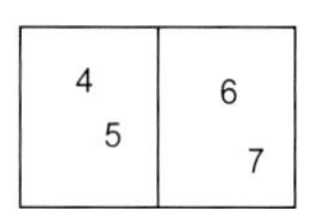

4. 购物中心内部
Mall interior
5. 项目区位图
Locatio map
6. 场地条件响应策略
Site analysis
7. 历史剧院
Historical theatre compound

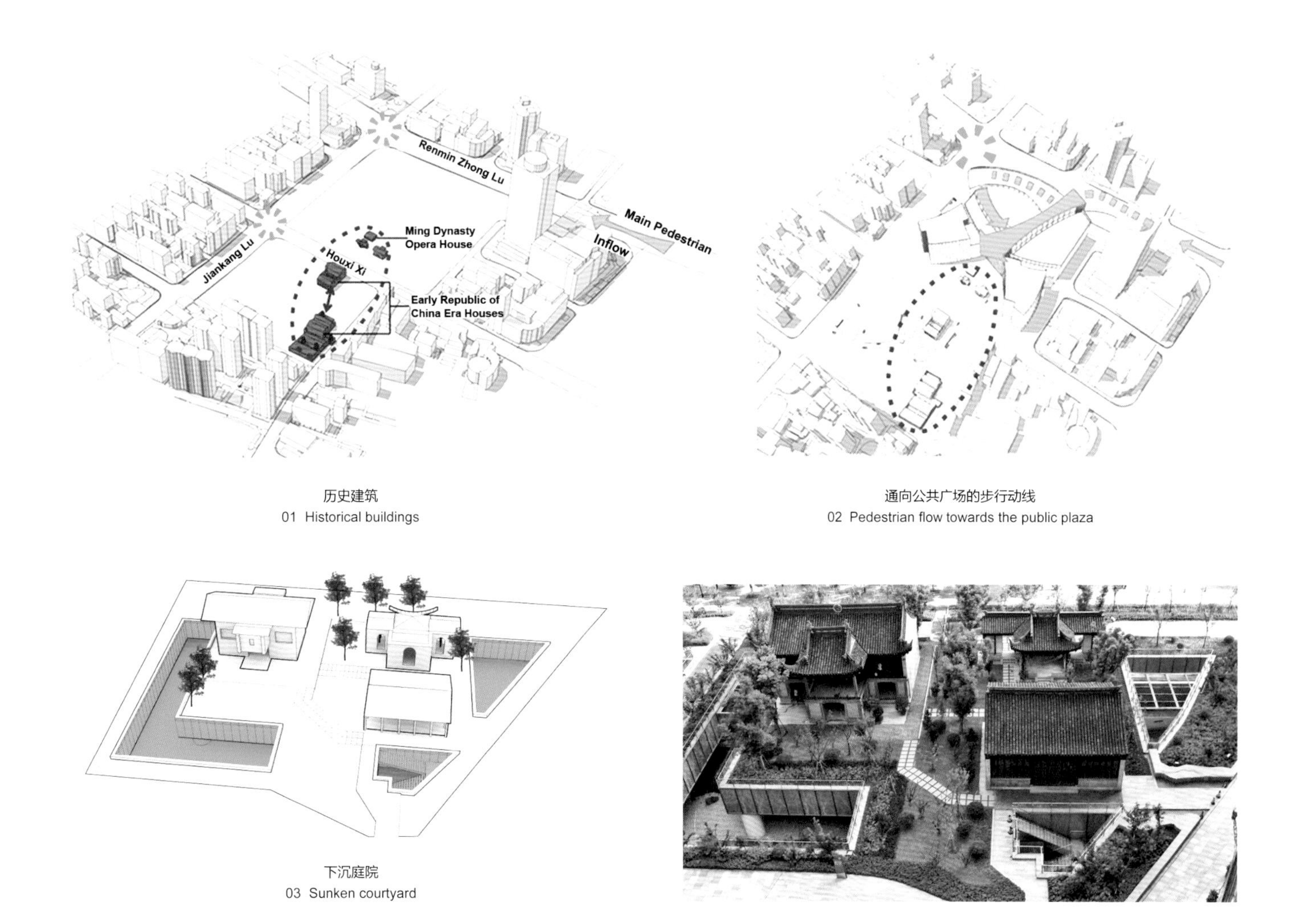

历史建筑
01 Historical buildings

通向公共广场的步行动线
02 Pedestrian flow towards the public plaza

下沉庭院
03 Sunken courtyard

a convenient point of access. Throughout the development, buildings of different eras are engaged in a timeless dialogue, creating an interesting timeline in space.

Creating a harmony between old and new was the starting point of the project. Architects created a dialogue between the historical buildings and the new buildings by drawing inspiration from the essence of Chinese calligraphy, with bold and smoothly curved lines that strengthen the form and space of the project. The form of the office building consists of two narrow boxes connected shoulder to shoulder with the curtain wall of the main facade ascending diagonally from bottom to top. The six-storey-high retail podium features a dazzling and complicated glass facade instead of normal curtain wall, forming a smooth full-glass surface. The planning of the project weaves together different functions, spaces and destinations. The entrance of the office building serves as a portal that extends the interior circulation to the outdoor plaza. The project resculpted civic space in keeping the spirit of the times while also treasuring the history of the city.

BALLY

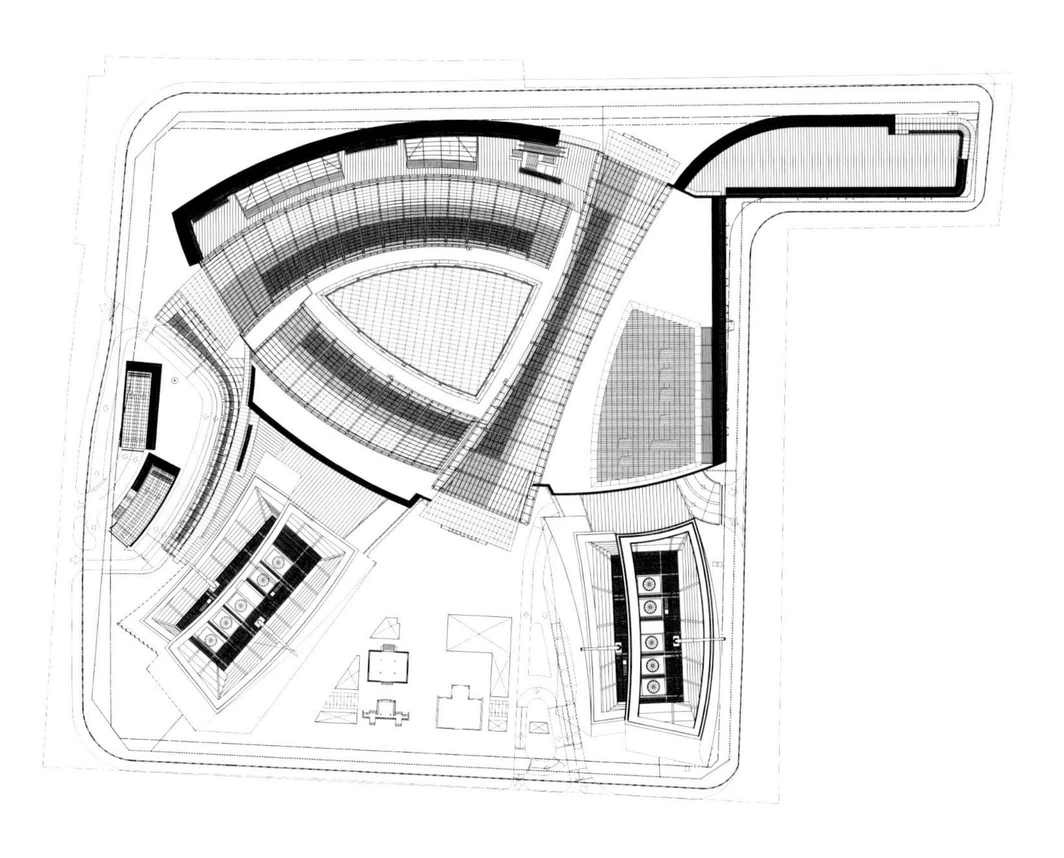

8	9 10 11

8. 购物中心内部
Mall interior
9. 一期总平面图
General plan (phase I)
10. 南侧立面图
South elevation
11. 剖面图
Section

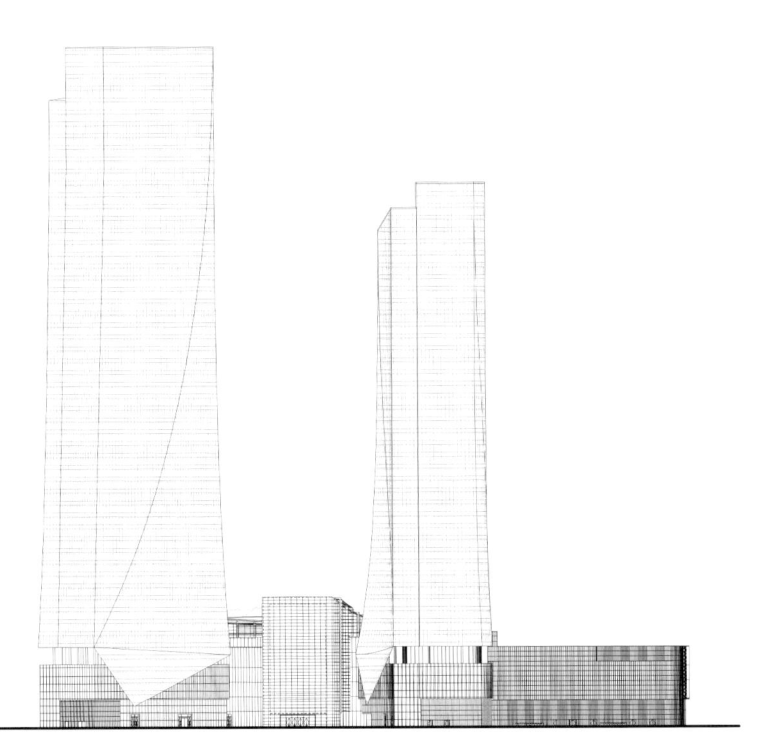

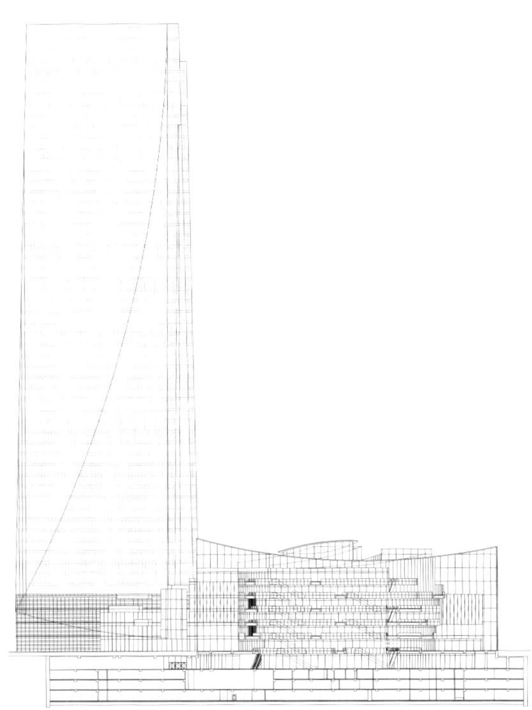

苏州西交利物浦大学中心楼

Xi'an Jiaotong-Liverpool University Central Building, Suzhou

城事：姑苏城外，太湖新石

苏州市旧称姑苏、平江，位于长江三角洲和太湖平原的中心地带，是著名的鱼米之乡、历史文化名城。几千年来，苏州不仅保留下“水陆平行，河街相邻”的双棋盘城市规划格局和“粉墙黛瓦、小桥流水”的古朴风貌，而且业已发展成为一个与时俱进的现代化城市。西交利物浦大学的中心楼便坐落于古城东侧的苏州工业园区内，这里风景秀丽、人文气息浓厚。中心楼伫立在校园北区的西侧，紧邻松涛街与文景路，是校内行政办公与信息图书馆藏的枢纽所在。这座建筑的设计灵感源于我国四大名石之一的太湖石。白居易曾说：“石有族聚，太湖为甲”，众多文人墨客也为太湖石留下诸多诗赞和佳语，太湖石也因此被称为“文人石”“学者石”，将太湖石作为教育办公建筑的主题原型更有其象征意义。设计师将其“瘦、皱、漏、透”的玲珑趣味和苏州古典园林的情致雅韵相结合，演绎出层次丰富、古今并蓄的建筑空间。大楼立面犹如层岩切片般的疏密分割，象征着知识的独立与整合，而其内部空间贯穿交错的处理手法，则呈现出一派别有洞天的景致。

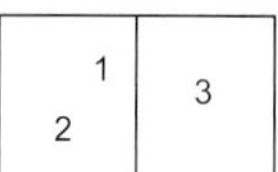

1. 苏州古典园林的一石一木是设计师的灵感来源
Classic Suzhou garden is the inspiration of the design
2. 西交利物浦大学中心楼全景
External view
3. 六层平台花园
Terrace on sixth floor

项目信息 | PROJECT INFORMATION

地　　点：中国苏州
建筑面积：59 893m²
设计时间：2008—2012 年
建成时间：2013 年
项目功能：教育设施
业　　主：苏州工业园区教育发展投资有限公司
主要设计人：温子先博士

Location: Suzhou, PRC
Gross floor area: 59,893 m²
Design time: 2008-2012
Completion year: 2013
Sector: Education
Client: Suzhou Industrial Park Education Investment Development Co. Ltd.
Director: Dr Andy WEN

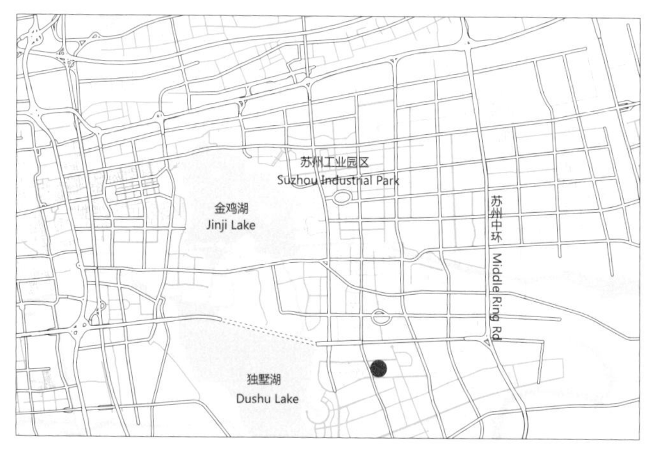

造型：外正内曲，道法自然

中心楼好像是一尊无比例的雕塑艺术品，不分大小、遑论高矮。仰视时呈正方体，代表着学术研究的严谨与方正；待到俯瞰观察，又像自由徜徉、散落在大地上的“魔方”。建筑的立面被无数尺度不同的穿孔铝板细致分割，疏朗之间可见许多结构精巧的“缝隙”，将楼梯和窗户完美融合在外立面内。步入 6 楼的中庭花园，会讶异地发现大楼内部的曲折回环与外部的方正姿态迥然不同。正如久经流水打磨抛光的太湖石，中心楼内部也被孔洞穿凿连结，其虚实空间的位置毫不矫揉，宛如自然之笔泼墨挥毫，一气绘就的天然画卷。

功能：多元融合，灵动布局

作为大学的主楼，它为师生提供了多种的独立功能，其中包括行政中心、图书馆、信息技术培训和学生活动中心等。不同功能的空间看似各自为政、分区划分泾渭分明，却为使用者创造出产生交集与对话的机会：6 楼的中庭花园连通了阅览区与行政区，如天井一般的半室外空间供师生自由洽谈休憩；9 层的空中连廊衔接东西两侧，增加了空间交往的便利；11 层与 12 层的办公区域中，均配置了多个可直达屋顶平台的窗口，教员在工作之余可以在园林般的室外空间放松心情。位于 2 层行政区的国际项目申请部，其走道净宽度达 5.5m，办公室之外靠窗一侧放置着舒适的座椅书桌，前来报名或面试的学员在正式谈话之前可以在该区域休憩：窗外是满栽花木的通高半室外庭院，低头是宁静祥和的人文微空间，抬眼是中心楼极具动感与韵律的外部造型……须臾间，便能感受到设计师共情的人文关怀。

City: Outside the City of Gusu, A New Taihu Stone

Suzhou, formerly known as Gusu or Pingjiang and located in the central zone of the Taihu Lake plain and Yangtze River Delta, is a famous "land of fish and rice" and a historical and cultural center of China. For a thousand years, Suzhou has not only preserved the double-grid urban fabric of "water parallel to land, road adjacent to stream" and the picturesque cityscape composed of white-walled, black-tiled dwellings and bridges over rivers, it has also developed into a modern city keeping pace with the times. The Central Building of Xi'an Jiaotong-Liverpool University is located in Suzhou Industrial Park to the east of the old city. The surrounding landscape is beautiful and it has a long cultural history. The Central Building,

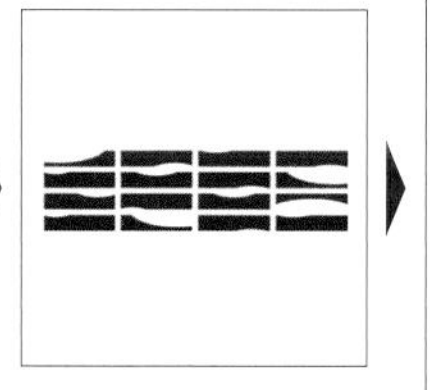

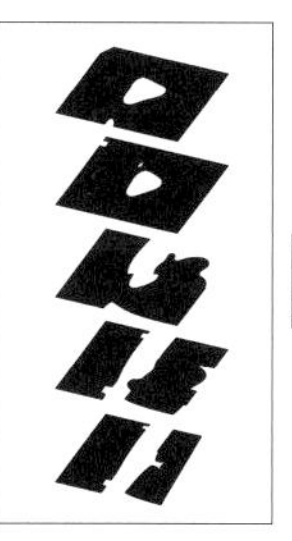

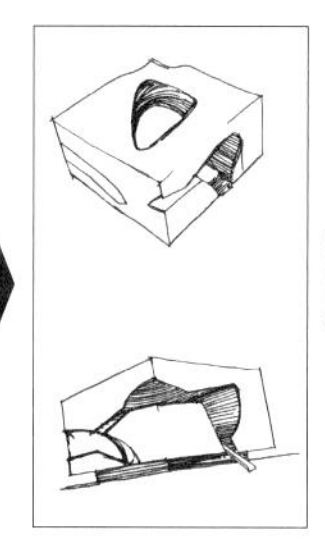

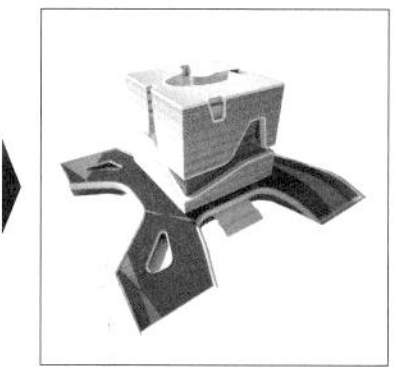

4	6
5	7

4. 西交利物浦大学中心楼夜景
Night view
5. 项目区位图
Location map
6. 概念推导及手绘图
Concept development and sketches
7. 中心楼局部外观
Partial view

built on the west side of the north campus, next to Songtao Street and Wenjing Road, is the university's center of administration and information. The design was inspired by the famous Taihu stones, one of the four great stones in China. Bai Juyi mentioned in *Tai Lake Stone Note* that "Stones being a family, Taihu as the best." Numerous other scholars have left many poems and verses for the Taihu stones, and thus Taihu stones are also known as "scholar's stones" and "literati's stones". Using it as the theme of the educational office building is particularly meaningful. Integrating the exquisite taste of the "thin, wrinkled, porous, transparent" Taihu stones and the atmosphere of a Suzhou classical garden, the architect designed a space with rich layers that contains both old and new. The facade of this building is like layered shale, symbolising the harmony between individual bits of historical knowledge, while the cross interlacing treatment of the interior creates a spectacular and unexpected view.

Form: Exterior Straight Versus Interior Curve, Imitation of Nature

The central building is like a sculpture blurring the sense of proportion — missing a sense of big or small, tall or low. It is a cube when you look up to it, and this symbolises the righteousness and preciseness of academic research. The facade of the building is covered with layers of perforated aluminium panels that each

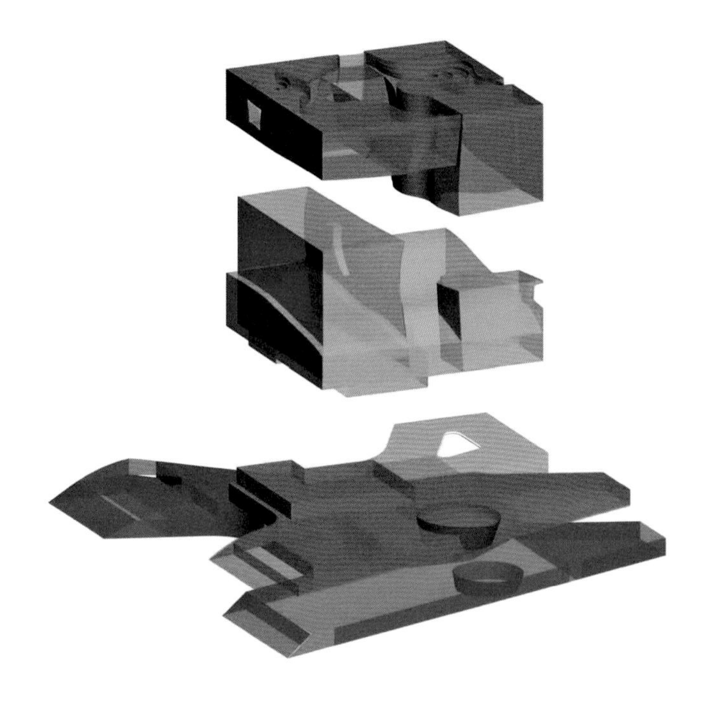

- 行政中心 Administration center
- 学生信息中心 Student resources center
- 培训中心 Training center
- 学生活动中心 Student activities center

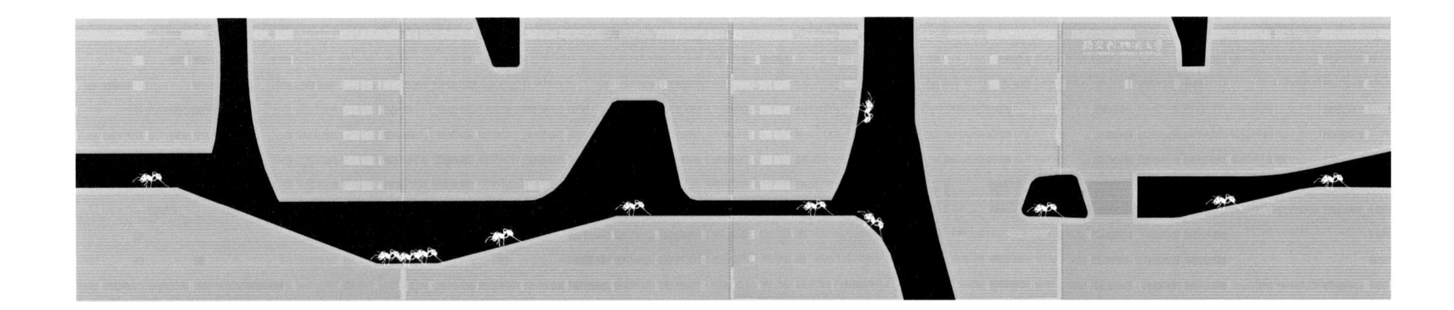

projects out at a different distance. The staircase and windows are exposed, becoming integral parts of the facade due to the gap between the aluminium panels. Standing in the atrium garden on the sixth floor, people will be surprised by the curving, flowing interior space, which is completely unlike the exterior. Just like the Taihu stones eroded and polished by flowing water, the interior of the building is porous and thus spatially continuous. The composition of solid and void is like a scroll painting made without stopping.

Function: Diverse and Flexible

As the central building of the university, this structure contains various independent functions for students and staff. The facilities include an administration center, library, information technology training and a student activities center. Spaces are specifically defined but at the same time allow intersection and communication. The courtyard on the sixth floor connects reading area and administrative area and the patio-like semi-open space provides free communication space for professors and students. A sky bridge on the ninth floor links up the east and west sides, increasing accessibility. Office areas on the 11th and 12th floors have windows directly connected to the rooftop, which gives faculty the chance to relax in a garden-like open space. On the second floor, near the international project application department, the width of the hallway is up to 5.5 meters so that there is ample space to put comfortable tables and chairs for students who come for applications and interviews. Outside the windows is a semi-outdoor courtyard filled with trees and plants. Downwards is a peaceful humane space; upwards is the dynamic, rhythmic exterior of the central building. It's a space that resonates with the sympathetic, humanistic care of the architect.

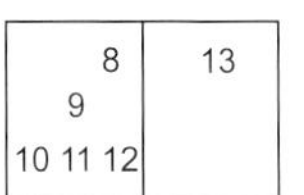

8. 功能空间分析图
Functional space analysis
9. 立面研究
Facade study
10. 外立面细部
Facade details
11.12. 镂空铝板的出挑尺寸都有细微的差别，以体现建筑立面湖石粗粝的自然感
Hollow aluminium plates of slightly different sizes give the building a texture reminiscent of Taihu stones
13. 中心楼仰视
Upward view

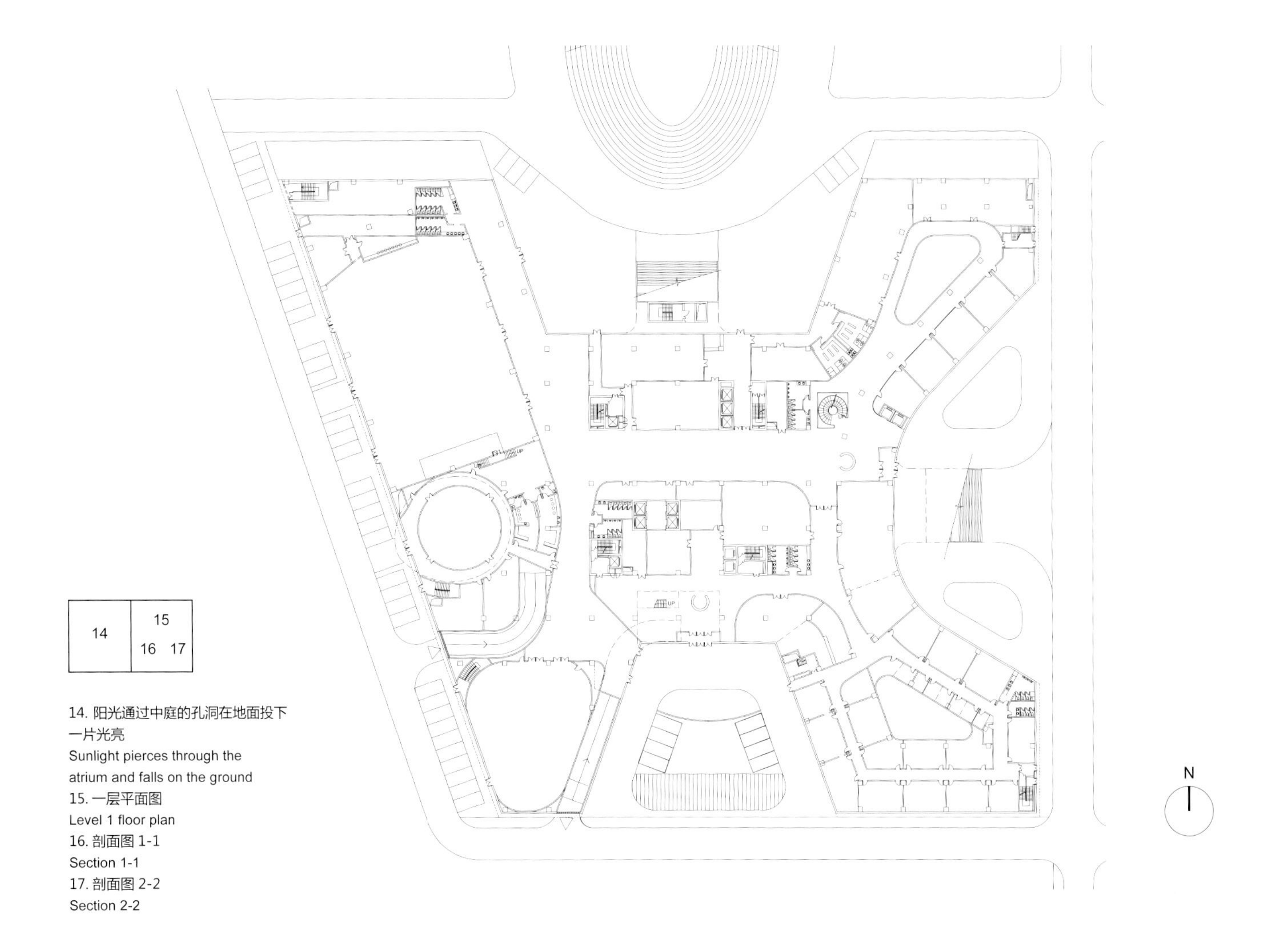

14. 阳光通过中庭的孔洞在地面投下一片光亮
Sunlight pierces through the atrium and falls on the ground
15. 一层平面图
Level 1 floor plan
16. 剖面图 1-1
Section 1-1
17. 剖面图 2-2
Section 2-2

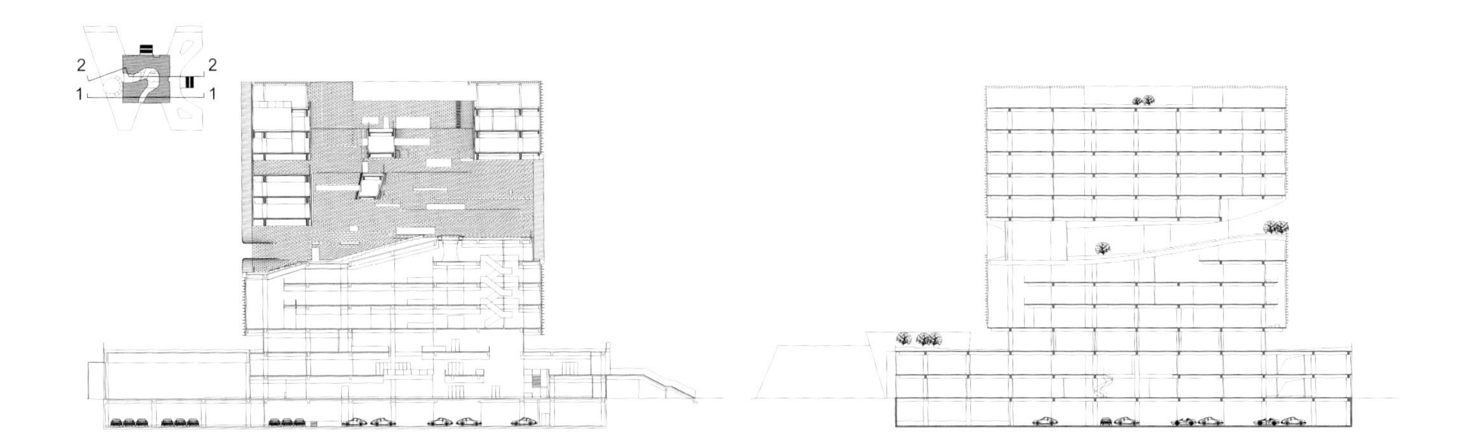

义乌之心
The Heart of Yiwu, Yiwu

义乌之心位于浙江省义乌市市工人路以北、北门街以西、新马路以东，坐拥义乌市商业中心——绣湖广场商圈。地块周边还有义乌市政府、绣湖公园、市民广场等场所。

项目涵盖商业零售、休闲娱乐、餐饮美食、生活服务、文化演艺、展示体验等多种业态，包括大型购物中心、景观广场、室内外步行街和地下停车库等设施，项目旨在将义乌之心打造成为：一站式城市生活广场、体验休闲文化的购物中心、主题性的商业综合体、前瞻性的国际化商业平台、国际化的城市形象展示窗口和现代性的文化演艺场所。

设计将义乌传统的城墙与充满现代感的新建筑融为一体，利用现代建筑手法创造兼具传统和现代生活文化的都市综合体。绵延的灰砖、大片的实墙，经典城墙元素再现了义乌拱辰门的恢宏气势。错落有致、从有至无的窗洞变化寓意着义乌的发展历程。漫步建筑位于北门街一侧，仿佛穿梭时光，置身于 20 世纪 20 年代的义乌；而行至康园路或是进入建筑内部，则回到了时尚前卫的当代。

建筑通过体块的穿插与交错，创造出层层退台及高低错落的台阶式户外平台，打造出标志性商业旗舰形象的同时，又与义乌北门的四十二台阶相呼应。建筑承载着历史，又蕴含了崭新的城市活力。

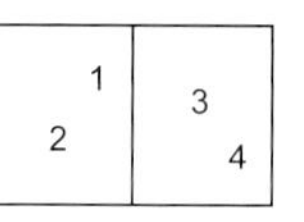

1. 充满历史感的义乌传统城墙亦融入了设计理念之中
Yiwu's old city wall, which has been reflected in the design
2. 体块的穿插与交错
Interplay of masses
3. 通透的入口体块
Entrance
4. 项目区位图
Location map

项目信息 | PROJECT INFORMATION

地　　点：中国义乌
建筑面积：146 000m^2
建成时间：2017 年
项目功能：商业零售
业　　主：杭州商旅投资发展有限公司
主要设计人：刘程辉，纪达夫

Location: Yiwu, China,
Gross floor area: 146,000 m^2
Completion year: 2017
Sector: Retail
Client: Hangzhou Commerce and Tourism Group Co., Ltd.
Directors: Leo LIU, Keith GRIFFITHS

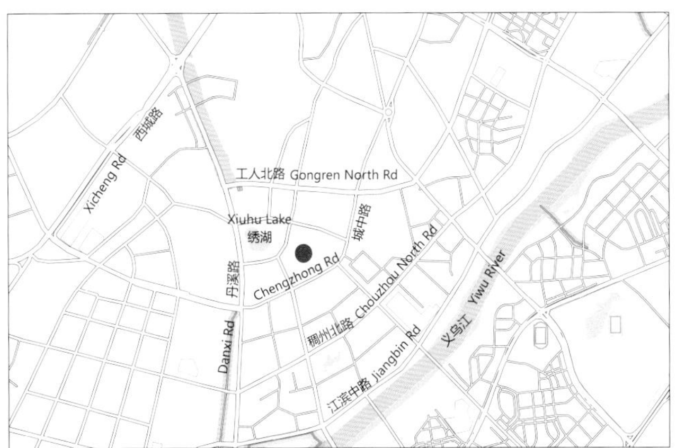

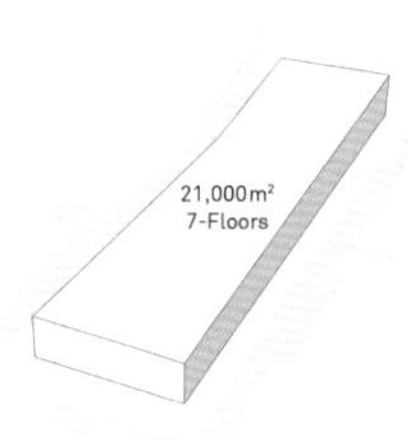

1. 体量
27 000 m^2 的商业空间分布于 7 层，以满足 50m 的项目限高

1. Area Massing
Seven floors of commercial space is needed to meet the 27,000 m^2 GFA requirements whilst maintaining within the 50m height limit of the site

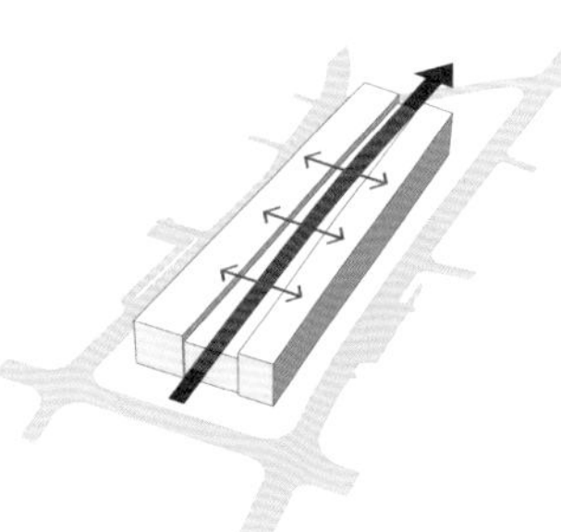

2. 商业主动线
体块分割为三个部分，以单一动线轴贯穿整个项目

2. The Main Commercial Circulation
The block is divided into three parts with a single circulation spine which penetrates through the massing

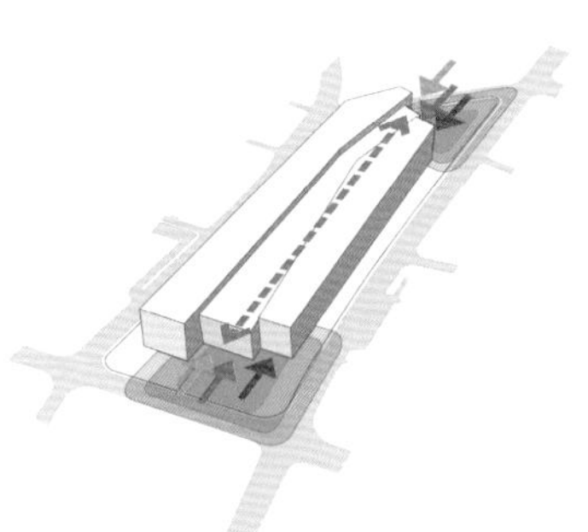

3. 入口及公共广场
主入口位于动线主轴的两端，各拥有一座大型室外公共广场，为项目入口带来活力

3. Entrance & Public Plaza
Main entrances are located at each end of the circulation spine, each with a large outdoor public plaza to activate the entrance

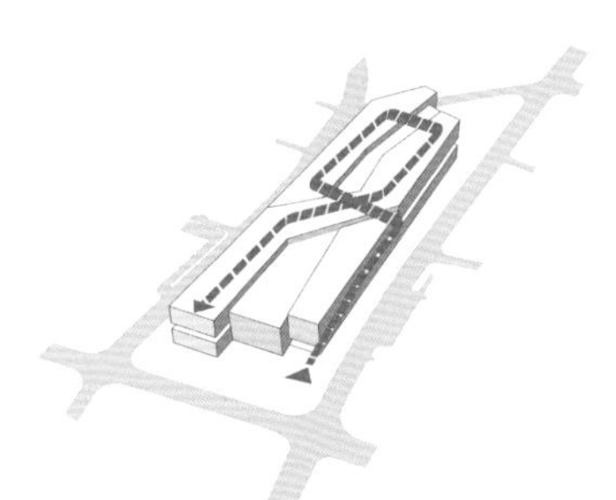

4. 24 小时全天候商业动线
用于餐饮、娱乐、展览等各类活动空间保证中央主轴线全天候的商业活力

4. 24-hour Commercial Circulation
Areas for dining, entertainment and exhibition are activated through the introduction of a 24-hour commercial loop bridging the central circulation spine

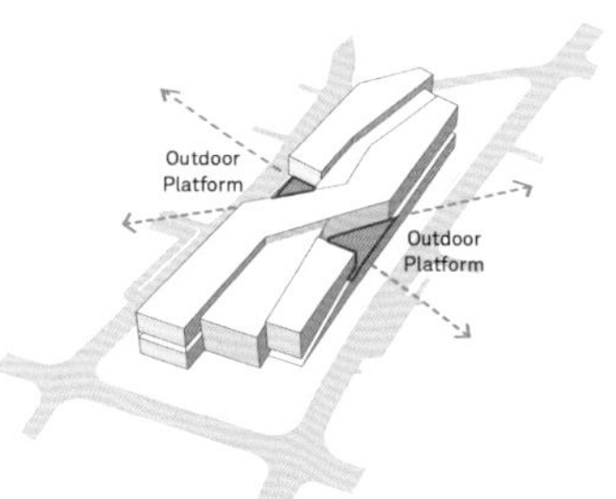

5. 户外露台
五层屋顶的户外公共露台，进一步加强 24 小时商业循环动线

5. Outdoor Platform
Introduction of outdoor public plazas on the fifth floor terrace further activates the 24-hour commercial loop

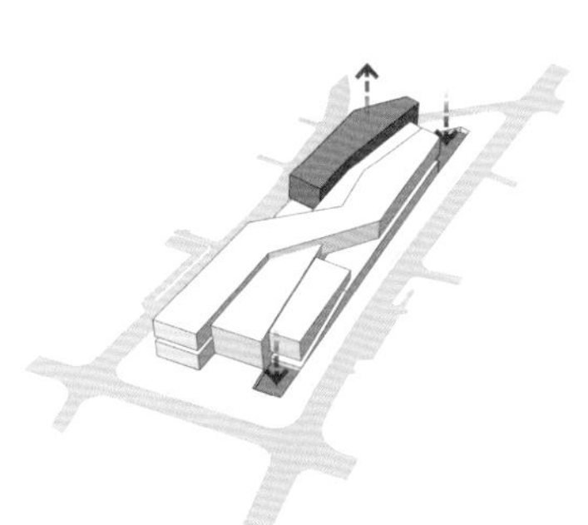

6. 体量优化
设计增加了屋顶的文化大厅和通往地下商业零售的下沉广场，进一步增加了整个项目的公共互动性

6. Refined Massing
The massing is completed though the expression of the cultural hall on the rooftop and the introduction of sunken plazas that connect basement retail to the ground floor, further increasing public interaction throughout the development

The Heart of Yiwu, located on the north of Gongren Road, west of Beimen Street and east of Xinma Road, is sitting in Xiuhu Plaza, the business center of Yiwu. The Yiwu city government, Xiuhu Park and a civic square are all located near the site.

The project contains retail, entertainment, dining, lifestyle destinations, cultural facilities and exhibition spaces, including a large-scale shopping mall, landscape square, an indoor/outdoor pedestrian street and underground parking facilities. It aims to offer multiple experiences to visitors by integrating lifestyle, shopping, food and beverage, cultural and leisure activities into one destination.

The design reimagines the traditional Yiwu city wall as part of the modern building, creating an urban complex with both traditional and modern influences. The grand atmosphere of the historic Yiwu Gongchen Gate has been revived with extensive use of grey tiles, large solid walls and other typical city wall elements. The subtle variations of the openings bring to mind the historical development of Yiwu into one of the world's most important trading hubs. Walking along Beimen Street, it is as if people are sent back to the 1920s, whereas walking along Kangyuan Road or inside the building brings people back to the modern age.

By shifting the building mass into interlocking pieces, Aedas has created layers of terraces and stepped outdoor platforms, which on one hand creates a unique flagship commercial complex, and on the other hand echoes the famous 42 steps of the North Gate of Yiwu. The building embodies both history and new vitality.

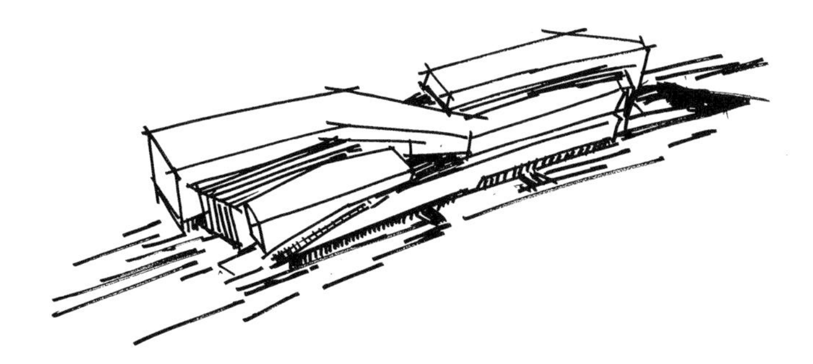

5 6	7 8 9

5. 内部公共空间
Interior of the shopping center
6. 形态演变
Development of form
7. 一层平面图
Level 1 floor plan
8. 立面图
Elevation
9. 手绘图
Sketch

广州南丰商业、酒店及展览综合大楼

Nanfung Commercial, Hospitality and Exhibition Complex, Guangzhou

南丰新展览综合大楼位于广州琶洲岛，未来该区域将被打造成为广州新的中央商务核心区。为契合该地区的发展定位， 又能从周边同类项目中脱颖而出，设计师致力于将建筑塑造为具有高度识别性的综合体。基地周边现有的展览建筑外墙采用卷边做法，为了与之明显区分，设计师在本项目中采用了直线而非流线的设计语言，并确定了“大型底座”和“铅笔盒”的体量组合模式。

该项目底部空间供展览使用，其上两座大楼高度相同，功能分别为商业办公与豪华酒店。两者具有共同的外观特征：四层或五层大型盒状结构承托起上方一系列不同长度、随意组合的细长盒状体量。相同的处理手法、不同的组合节奏使两栋相距 160m 的建筑既各具特色又风格统一。商业办公楼高 123m，尽管外形犹如随意堆叠而成，但内部却拥有充足的标准办公空间。15 层的豪华酒店客房沿大型中庭长边布局，电梯、消防楼梯及其他服务设施设置在两端。最引人注目的是大堂内部的重型对角支柱，在承托起上部楼层重量的结构的同时，塑造出令人震撼的空间效果。

1	3
2	

1. 广州城市风光
Cityscape of Guangzhou
2. 相同的细长盒状的处理方式使两栋建筑具有相似的外形，增强了整体感
The same design language ensures unity of the two buildings
3. 设计师采用直线的设计语言，使建筑具有独特的韵律感
The linear design language gives a special rhythm to the buildings

LANGHAM PLACE

项目信息 | PROJECT INFORMATION

地　　点：中国广州
建筑面积：159 000m²
设计时间：2007—2013 年
建成时间：2013 年
项目功能：综合体
业　　主：南丰集团
主要设计人：Andrew BROMBERG

Location: Guangzhou, PRC
Gross floor area: 159,000 m²
Design time: 2007-2013
Completion year: 2013
Sector: Mixed-use
Client: Nan Fung Group
Design Director: Andrew BROMBERG

The Nanfung Commercial, Hospitality and Exhibition Complex is located on Pazhou Island in Guangzhou, which is slated to become a new central business district for Guangzhou. In response to the area's planned development, the complex was designed with a strong identity to stand out from the surrounding buildings. In order to distinguish itself from the existing exhibition buildings nearby which have curving lines, the architects used rectilinear lines and decided that the building mass would take the form as a large podium with "pencil case" towers.

4	6
5	7

4. 酒店露台景观
Landscaped deck
5. 项目区位图
Location map
6. 酒店中庭
Atrium of hotel
7. 大堂内引人注目的重型对角支柱
Hotel lobby with eye-catching structure

The podium is mainly used for exhibition, while two towers of identical height above are for commercial offices and a luxury hotel, respectively. The two blocks share a similar appearance, with a four/five-storey-high podium topped by multi-storey box-shaped masses. The same language spoken with a different rhythm allows the two blocks, 160 meters apart from each other, to have their own identity while sharing the same style. The 123-metre office tower has ample standard office space despite the seemingly randomly stacked boxes that define its form. The nearby 15-storey luxury hotel positions its rooms along the long sides of the huge atrium while elevators, fire escape staircases and other services are on the two short ends. The most eye-catching structure is the diagonal support in the lobby which bears the load of upper floors, creating a striking space.

8	9
	10

8. 酒店大堂
Hotel lobby
9. 建筑外观
External view
10. 剖面图
Section

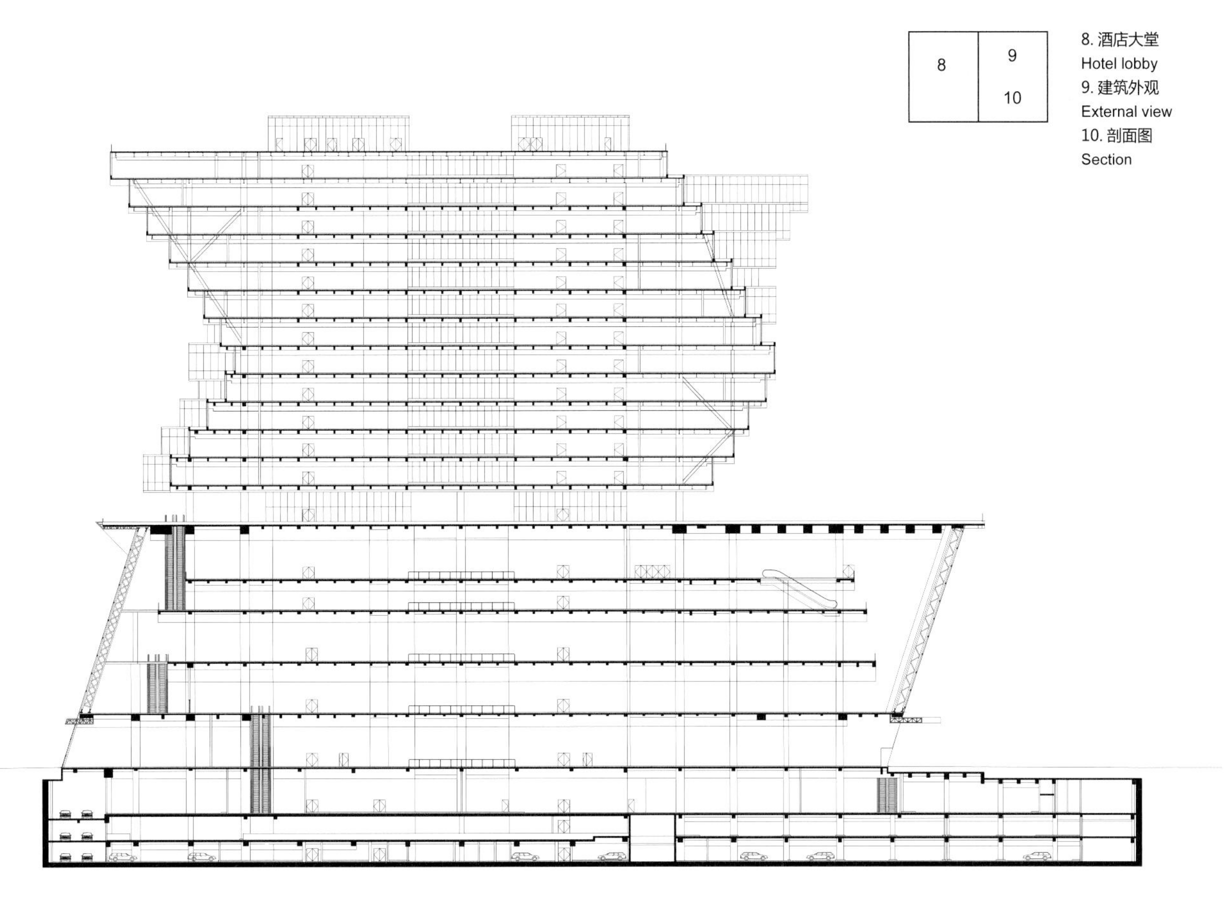

广州邦华环球贸易中心

Bravo PARK PLACE, Guangzhou

广州，位于珠江三角洲北缘，是中国第三大城市。自秦朝开始，广州一直是华南地区的政治、军事、经济、文化和科教中心，有“千年商都”之称。邦华环球贸易中心便位于这座繁华、包容的国际大都市。项目坐落于广州海珠区新港东路，琶洲互联网总部经济区的中心位置，集超甲级写字楼、五星级万丽酒店、莎玛服务式公寓、特色园林等于一体，以 230m 的高度绘就城市的新天际线，成为该区域最独特的地标建筑。

项目以“拼图”为设计理念，体现出容纳百川、包容并蓄的气魄和雄心，自然融入广州城市发展的脉动之中。项目整体分为两大部分，其中高层空间用于商业办公及服务式公寓，坐拥广州塔和珠江的绝佳景观，低层空间则为餐厅和精品酒店。每个建筑体块都拥有独立的核心筒和电梯，使不同类型的商业运营更具有灵活性。量身定制的综合体不仅能满足业主的各种要求，也能在设计上取得协调融合。建筑体块在统一于整体风格的同时，以错落而均衡的形态反映出内部空间的不同功能，展现出体块间平等友好的姿态。

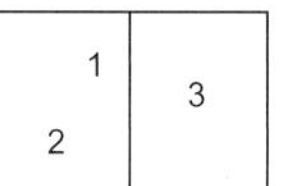

1. 广州作为中国南部的商贸中心和综合交通枢纽，是中国的“南大门”
Guangzhou is a trade center and integrated transportation hub in southern China. It serves as the “south gate” of the nation
2. 华灯初上的邦华环球贸易中心
Night view
3. 阳光下的建筑立面
The building facade in the sun

ABC
中国农业银

项目信息 | PROJECT INFORMATION

地　　点：中国广州
建筑面积：99 429m²
设计时间：2012 年
建成时间：2017 年
项目功能：综合体
业　　主：广东邦华集团
主要设计人：韦业启

Location: Guangzhou, PRC
Gross floor area: 99,429 m²
Design time: 2012
Completion year: 2017
Sector: Mixed-use
Client: Bravo Group
Director: Ken WAI

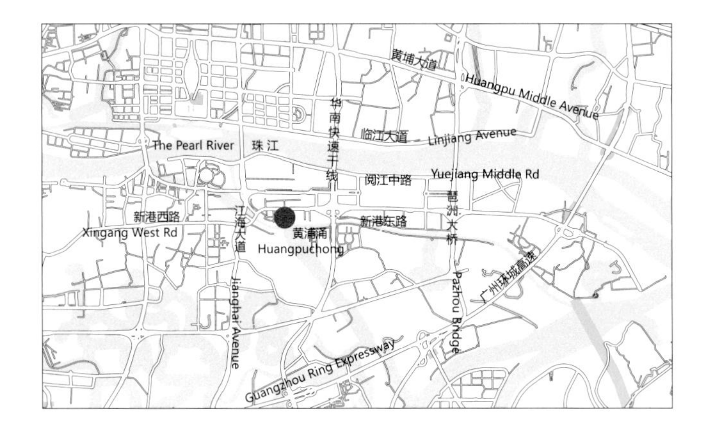

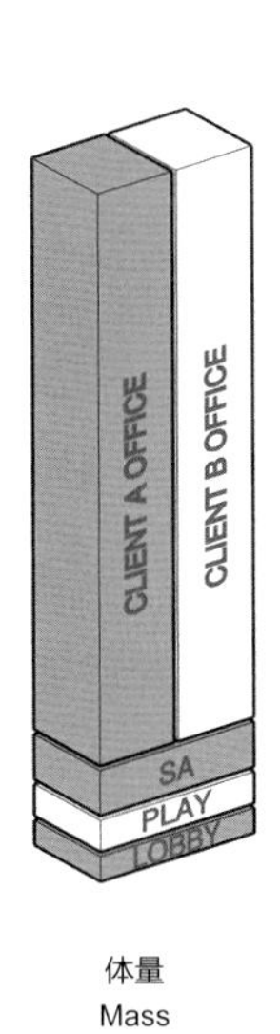

体量
Mass

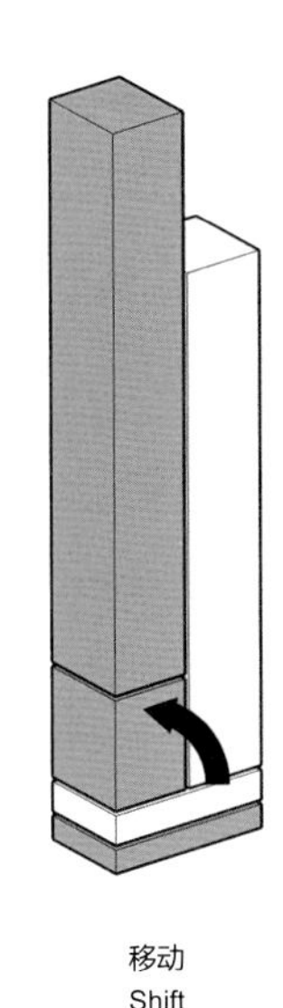

移动
Shift

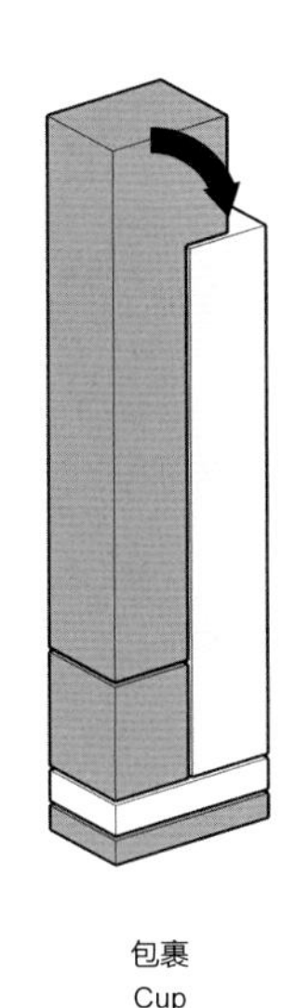

包裹
Cup

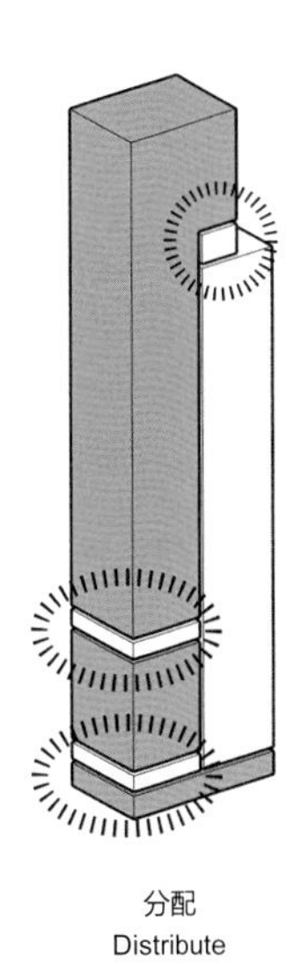

分配
Distribute

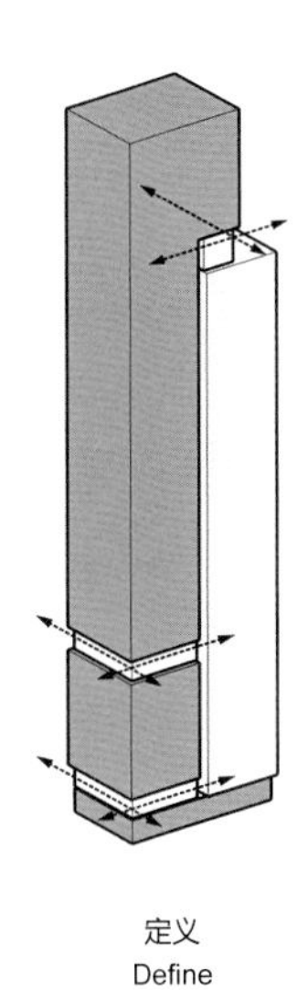

定义
Define

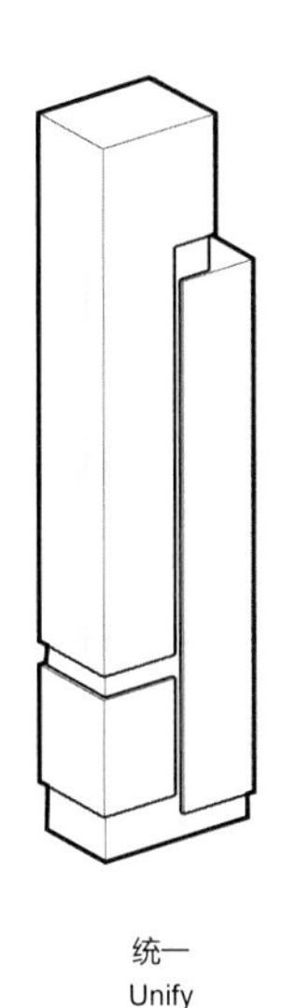

统一
Unify

4. 邦华环球贸易中心立面细部
Facade details
5. 项目区位图
Location map
6. 体块分析图
Mass analysis diagrams
7. 邦华环球贸易中心与远处的广州塔
External view with the Canton Tower as background

Guangzhou, located at the top end of the Pearl River Delta, is the third largest city in China. Ever since the Qin Dynasty, Guangzhou has been a political, military, economic, cultural and educational center for south China, known by its nickname, the “millennium commercial capital.” Bravo PARK PLACE is located in this flourishing international megacity. The project is set in the center of the Pazhou internet headquarters economic district on Xingang East Road in Haizhu District. It contains Grade-A offices, a five-star Renaissance Hotel, Shama Serviced Apartments, a unique garden and high-end dining. The 230-meter tower creates a new city skyline, becoming the most distinctive landmark in the region.

The design is inspired by a jigsaw puzzle, embodying the ambition and aspiration represented by the rivers of the Pearl River Delta, along with the inclusiveness that defines the pulse of Guangzhou's urban development. Bravo PARK PLACE is a tower combining commercial offices and serviced apartments on the upper floors, enjoying spectacular views of both Guangzhou Tower and the Zhujiang River, with restaurants and a boutique hotel at the lower levels. Both parts have a separate core and elevators, maximising the commercial flexibility. The tailor-made complex can not only fulfill different client requirements, it also manages to integrate everything into one seamless design. The distinction between the functions is expressed on the facade, which is unified by a semi-controlled fenestration that uses vertical fins at regular intervals.

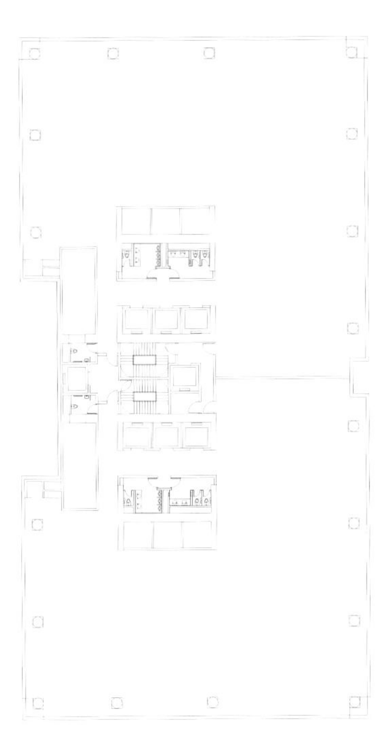

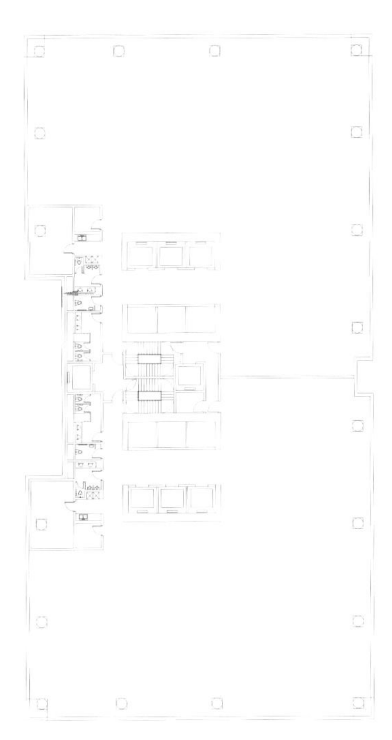

8	9 10 11
	12

8. 远观邦华环球贸易中心
External view
9.10.11. 标准层平面图
Typical floor plans
12. 大堂内部
The lobby

珠海粤澳合作中医药科技产业园总部大楼

Headquarters, Traditional Chinese Medicine Science and Technology Industrial Park of Co-operation between Guangdong and Macao, Zhuhai

珠海横琴位于珠江口西侧，毗邻港澳，四面环水，海湾众多，与澳门特区遥相对望，是一个具有开放活力的智能生态岛。珠海粤澳合作中医药科技产业园总部大楼位于横琴西北角，是集办公、服务、会展、商业孵化为一体的综合办公建筑。建筑充分体现了珠海横琴“合作、创新、服务”的发展理念，发挥城市地域优势，推进港澳紧密合作、融合发展。

主楼由两座塔楼组成。西楼外形方正，围绕着中庭，有供人们休憩的室外空间。东楼则缓缓环绕于双中庭，与西侧楼形成对位。建筑设计灵感来源于中国“天圆地方”的思想，展现了人类社会与自然之间和谐共生的关系。与传统的办公楼设计不同，珠海粤澳合作中医药科技产业园总部大楼设计了一个 15 层的“阳光房”，整座建筑由玻璃和铝材构建而成，从上往下分解成了一系列的“盒子”。建筑的平台呈现并创造出了一个松散的空间网格，为屋顶的户外花园营造了一个完美的环境。整栋建筑外形疏漏有致，自然被融入到了建筑之中，建筑与自然形成对话关系的同时，也能够与当地的气候相互呼应，并且被赋予了更多的功能和更丰富的体验。

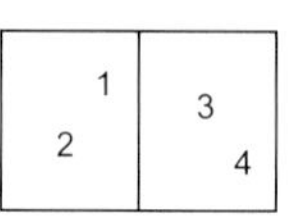

1. 珠海城市景观
Cityscape of Zhuhai
2. 建筑南立面，环北大道视角
South elevation facing Huandao North Avenue
3. 建筑东南立面，环北大道视角
Southeast elevation facing Huandao North Avenue
4. 项目区位图
Location map

项目信息 | PROJECT INFORMATION

地　　点：中国珠海
建筑面积：68 000m²
设计时间：2015 年
建成时间：2018 年
项目功能：办公楼
业　　主：粤澳中医药科技产业园开发有限公司
主要设计人：刘程辉，纪达夫

Location: Zhuhai, PRC
Gross floor area: 68,000 m²
Design time: 2015
Completion year: 2018
Sector: Office
Client: Guangdong-Macao Traditional Chinese Medicine Technology Industrial Park Development Co., Ltd.
Directors: Leo LIU, Keith GRIFFITHS

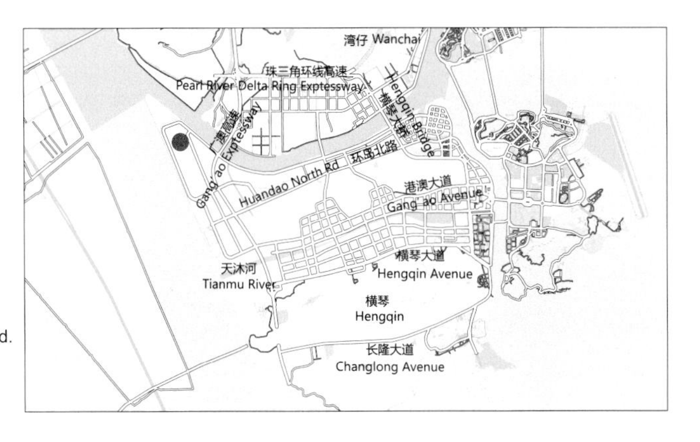

Hengqin Island sits on the west side of the Pearl River Estuary, close to Hong Kong and Macao, and it is an open, vital, intelligent and sustainable island. Located at the northwestern corner of Hengqin, this project is an integrated office building that accommodates offices, services, conference and exhibition spaces and business incubation spaces. The building embodies Hengqin's development concepts of cooperation, innovation and service, as well as Zhuhai's vision of taking advantage of its strategic location by promoting cooperation with Hong Kong and Macao.

The main buildings consist of two towers. The West Building is rectilinear, oriented around an atrium with a powerful deconstruction of the lower floors, creating shade, porosity and outdoor terraces. The East Building flows gently around its twin atriums in counterpoint to its neighbouring building. The architectural form is inspired by the Chinese concepts of "Round Sky and Square Earth" in ancient Chinese philosophical thinking. It shows the desire for a harmonious relationship between natural world and human society. Different with traditional office plan, this headquarter building has been designed with a 15-floor sunny atrium and an entire body breaking up into a series of boxes from the top down. The podium of the building has created a loose space network, which builds a perfect environment for the outdoor garden on rooftop, whilst being reflective to the local climate. The design creates not only an enjoyable form but also a rich spatial experience.

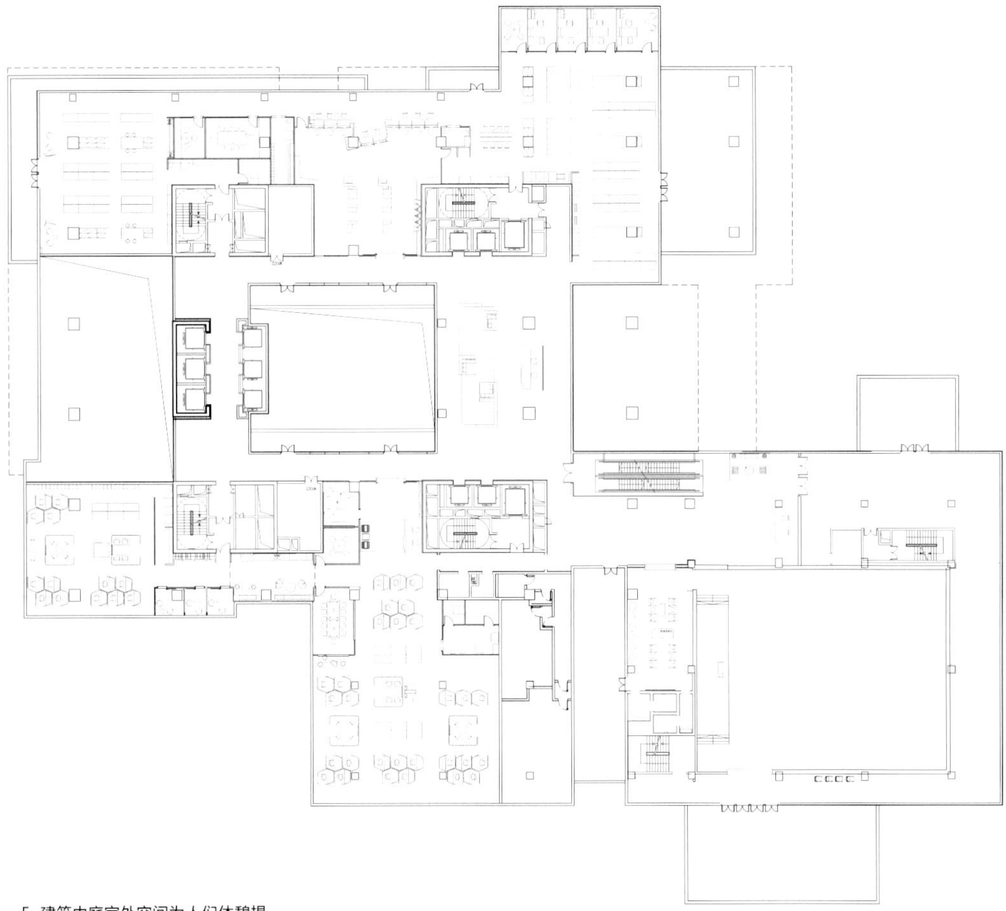

5	6
	7

5. 建筑中庭室外空间为人们休憩提供了绝佳去处
Atrium
6. 二层平面图
Level 2 floor plan
7. 建筑剖面图
Section

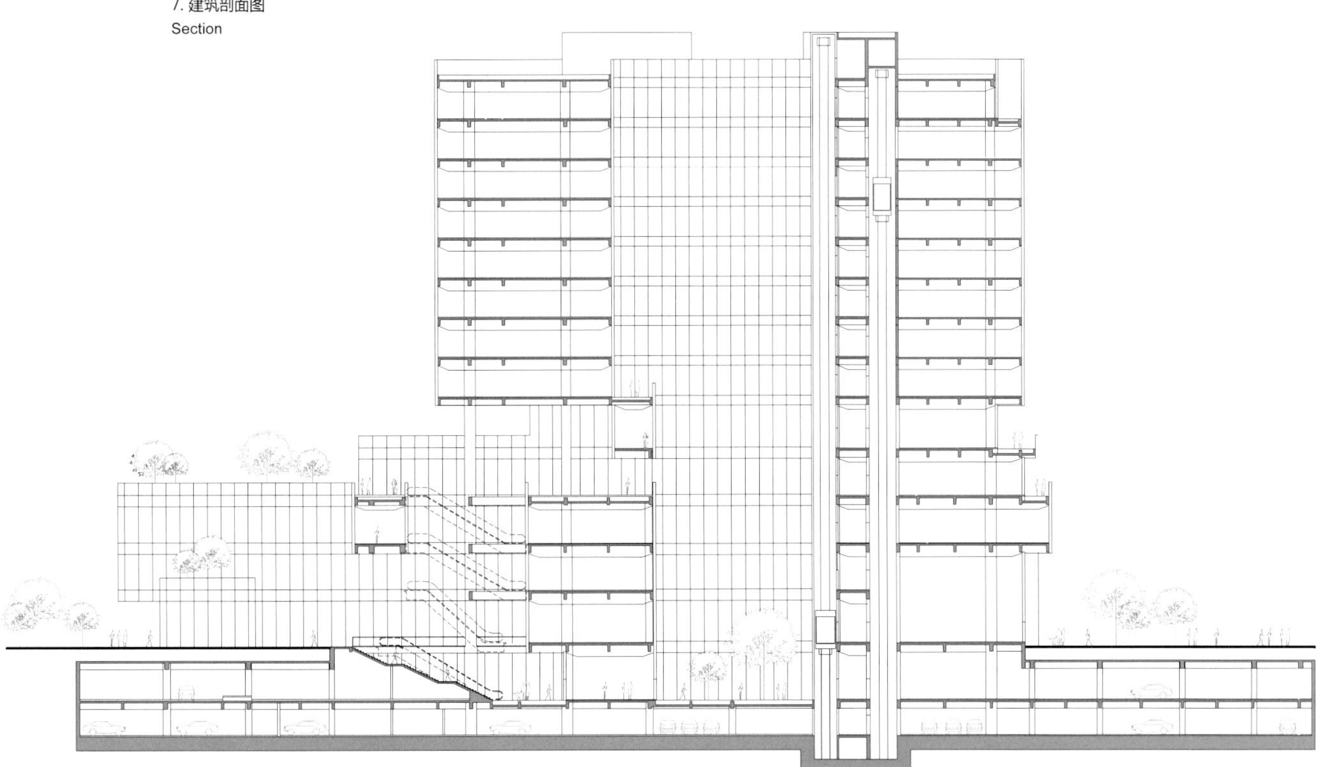

深圳宝安国际机场卫星厅

Shenzhen Airport Satellite Concourse, Shenzhen

深圳宝安国际机场位于珠江口东岸的滨海平原，距离深圳市区 32km，场地辽阔，净空条件优良，是一个具有海、陆、空联运功能的现代化航空港，是世界百强机场之一，担负着重要的空港运输任务。Aedas 赢得深圳宝安国际机场卫星厅设计的国际竞赛，进一步增强了其在基础设施方面的设计实力，并彰显 Aedas 在提供创新性航空解决方案上与日俱增的影响力。

设计以其梦幻且具未来感的建筑外观和室内空间延续了宝安机场 T3 航站楼的设计语言，并传达出极具活力的深圳城市精神。客运廊的屋顶设计灵感源于深圳当地蜿蜒的河流，这一设计理念强化了空间的流动感。旅客可乘坐自动捷运系统（APM）到达光照充足的客运廊中心。多层零售及就餐区域提供丰富多样的环境及活动空间，并自然引导乘客前往登机口。抵达旅客将被引导至建筑外缘，在去往自动捷运系统及主航站楼的行程中可以充分感受自然光照的魅力。

Aedas 作为主创建筑设计师，与广东省建筑设计研究院（GDAD）、航空规划方 Landrum & Brown 共同组成国际顶尖设计团队，共同打造出新深圳宝安国际机场卫星厅项目。

1. 深圳是一座极具活力的城市，机场设计以梦幻且具未来感的建筑外观和室内空间传达了深圳城市精神
The design of the airport concourse reflects the dynamics of the city with a futuristic architectural form and well-planned spaces
2. 俯瞰卫星厅
Aerial view of concourse
3. 俯瞰宝安机场
Aerial view of Shenzhen Bao'an International Airport
4. 项目区位图
Location map

项目信息 | PROJECT INFORMATION

地　　点：中国深圳
建筑面积：230 000m²
项目功能：基础设施
业　　主：深圳机场（集团）有限公司
主要设计人：江立文，唐宙行

Location: Shenzhen, PRC
Gross floor area: 230,000 m²
Sector: Infrastructure
Client: Shenzhen Airport Co. Ltd.
Directors: Max CONNOP, Albert TONG

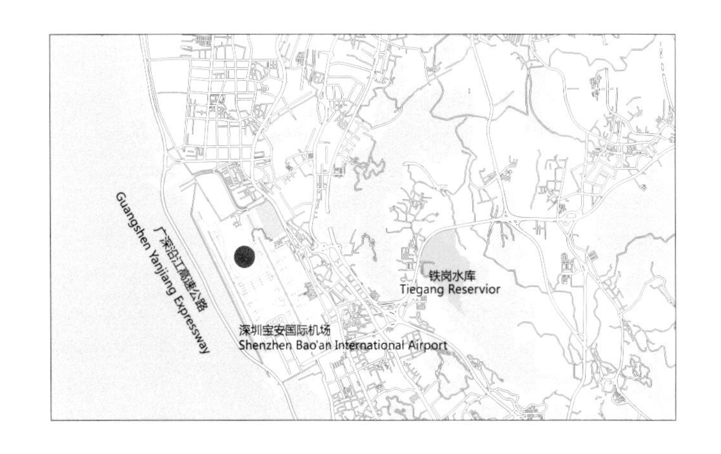

Shenzhen Bao'an International Airport is located on the east bank of the Pearl River, 32 kilometres away from the downtown area of Shenzhen. With a vast and open land area, the airport is considered a top-notch modern hub with land and sea connections; it is ranked as one of the top 100 airports in the world. Aedas won an international competition for the airport's new satellite concourse, further reinforcing Aedas' strength in infrastructure design, as well as demonstrating its increasing influence in the world of innovative aviation solutions.

The design of the concourse inherits the spirit of the design for Terminal 3, with surreal and futuristic appearance and space, which conveys the energetic spirit of Shenzhen, a young and fast-growing city. The roof design was inspired by the gently curving rivers of the region, reflecting the flow of the passengers through the concourse. Passengers arrive at the heart of the concourse that is filled with natural light from an automated people mover station. Multi-level retail and food and beverage areas provide a rich variety of environments and areas for passengers to experience, guiding them intuitively towards the departure gates. Arriving passengers are directed to the perimeter of the building, allowing them to experience natural diffused daylight as they move toward the automated people mover and onwards to the main terminal.

Aedas is the lead design architect in an international team that consists of GDAD as the local design institute/terminal planners and Landrum & Brown as aviation planners.

5. 客运廊屋顶设计灵感源于深圳当地蜿蜒的河流
The roof form is inspired by the gently curving rivers of Shenzhen
6. 室内丰富多样的环境及活动空间
Diverse interior space
7. 室内可充分感受到充足光照
Dramatic effect of lighting

珠海横琴国际金融中心

Hengqin International Financial Center, Zhuhai

区位：超级大都会上的明珠

珠海，东与深圳、香港隔海相望，南与澳门陆地相连。随着港珠澳大桥的通车，珠海与香港、澳门紧密融合，成为内地唯一陆路与香港、澳门相连的城市，三地交通往来只需 30 分钟，更有言称珠海的发展潜力将超过深圳。横琴国际金融中心位于珠海十字门中央商务区横琴片区近海金融岛上。十字门中央商务区地处西江入海口，与澳门一水相隔，是珠江三角洲规划中的未来金融新区，由 Aedas 设计的横琴国际金融中心将成为珠海横琴岛的地标。

横琴国际金融中心 333m 高的办公及公寓塔楼能俯视整片水域，塔楼下方为带有商业零售和会议会展设施的裙房。来源于低处的巨大能量汇聚奔涌，喷薄而出，在城市高空中凝固成四束融为一体的塔楼建筑，它象征着横琴汇集了珠海、澳门、香港和深圳的城市精华，成为珠江口超级大都会上的一颗明珠。

1 2	3

1. 珠海城市景观
Cityscape of Zhuhai
2. 空中鸟瞰图
Aerial view
3. 建筑外观
External view

项目信息 | PROJECT INFORMATION

地　　点：中国珠海	Location: Zhuhai, PRC
建筑面积：138 158m²	Gross floor area: 138,158 m²
设计时间：2012 年至今	Design time: 2012 till now
建成时间：2019 年	Completion year: 2019
项目功能：办公楼	Sector: Office
业　　主：珠海十字门中央商务区建设控股有限公司	Client: Zhuhai Shizimen Central Business District Development Holdings Co,. Ltd.
主要设计人：温子先博士，纪达夫，林世杰	Directors: Dr. Andy WEN, Keith GRIFFITHS, Ed LAM

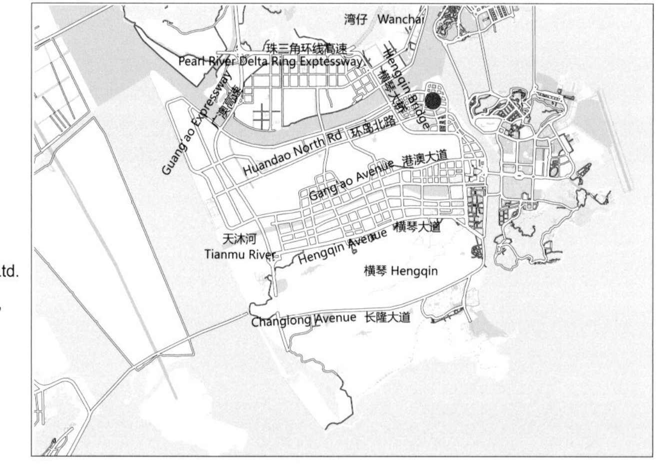

灵感：华夏龙文化精髓

华夏文明，以龙为图腾，横琴国际金融中心的设计灵感就来自中国古典绘画经典——南宋陈容《九龙图》中的造型，图中姿态各异的神龙在滔天骇浪、变幻风云中纵横穿梭，构图上虚实相映，有张有弛，气势惊人。横琴国际金融中心作为一座屹立在横琴岛中的地标建筑，从设计初始，建筑师们就着力于以建筑的语言来拓展地域特征的外延，以《九龙图》中的“蛟龙出海”为设计主题。在中国古典文学形象中，蛟龙是乘风破浪的神兽，通常用以形容新生力量，在“龙与海”的建筑造型互动中，也预示着未来这片金融新区的蓬勃发展。

造型：流线中的美学体验

横琴国际金融中心塔楼立面采用幕墙玻璃与金属面板，流泻而下，成为横琴随风飘动的巨大幕帘，不仅覆盖着裙房中高大的展厅，也成为极具识别性的入口雨棚。传统的高层建筑塔楼与裙房总是自成体量，缺少联动，Aedas 的建筑师们则将塔楼与裙楼接合处的直线条牵引、拉长、扭转，使裙楼的屋顶得以延展升腾为塔楼外墙，从而将高耸窄长的塔楼与相对宽阔低矮的裙楼完美结合在一起。

鉴于珠海每年台风季时间比较长的气候特征，横琴国际金融中心的流线造型顺应了这一气候特点，降低了塔楼的风荷载，也显得更轻盈动感。随着楼层增加，建筑平面从办公层的方形变化为酒店式公寓的 U 形，观景视野也随之扩展。同时，一座结合多项功能的大型特色中庭在高空中被 U 形公寓围合，为城市高空中的住户描绘出一幅云端的生活美学画卷。

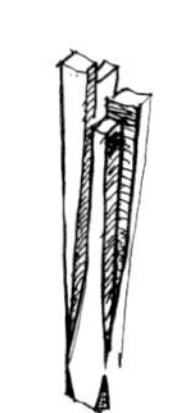

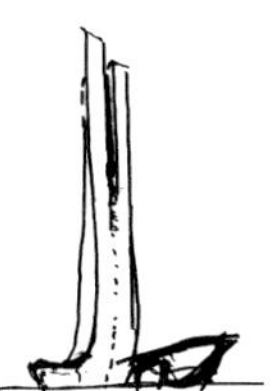

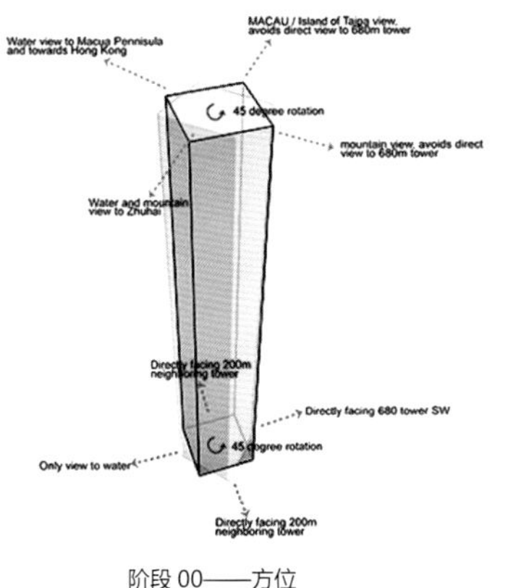

阶段 00——方位
Phase 00—Orientation

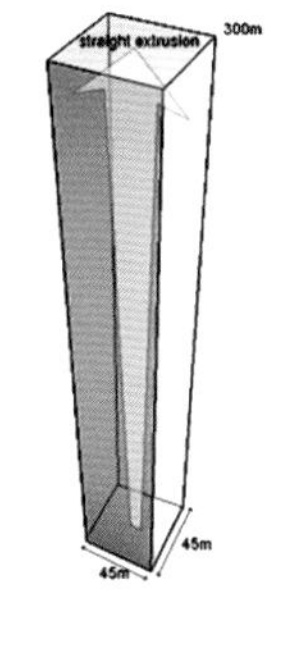

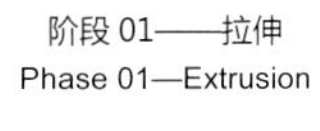

阶段 01——拉伸
Phase 01—Extrusion

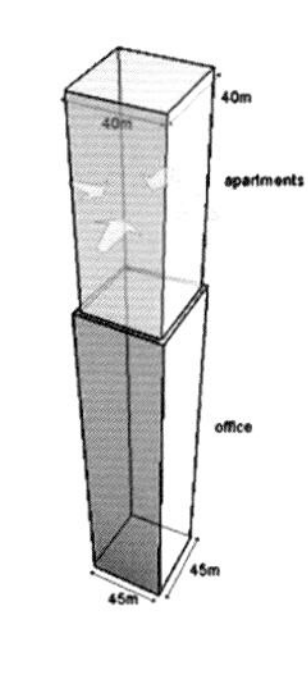

阶段 02——分割
Phase 02—Division

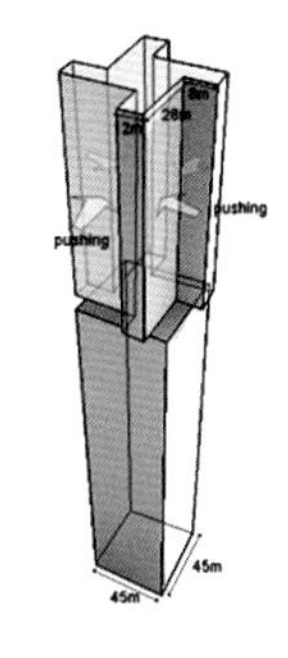

阶段 03——组合
Phase 03—Shaping

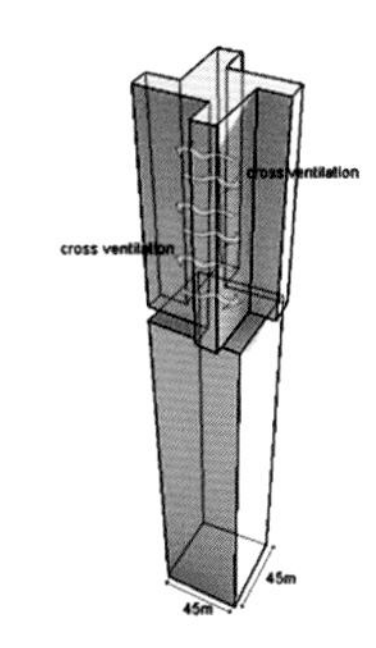

阶段 04——流通
Phase 04—Air

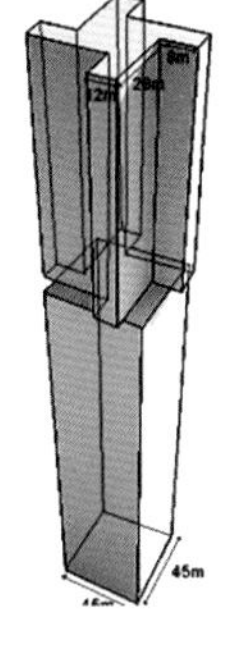

阶段 05——方位
Phase 05—Basic

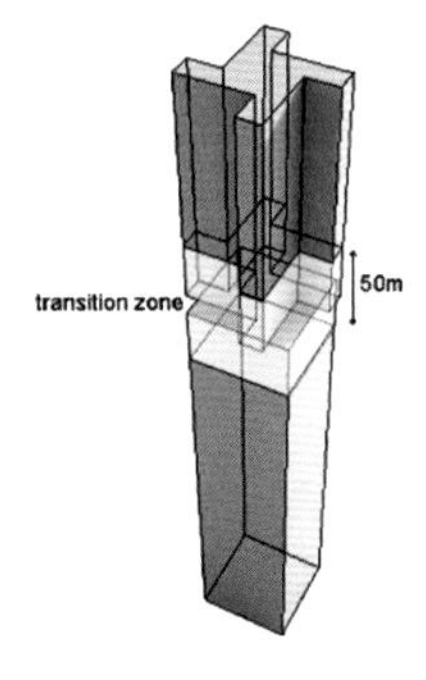

阶段 06——转换
Phase 06—Transition

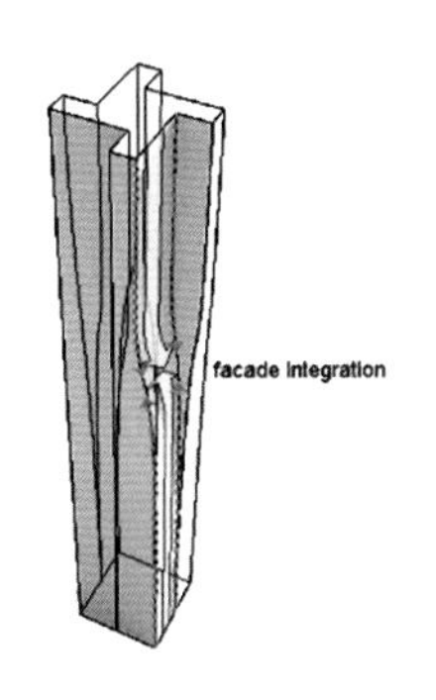

阶段 07——融合
Phase 07—Merging

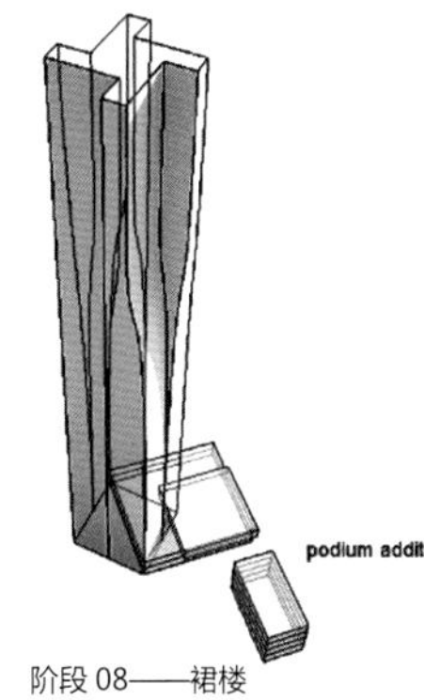

阶段 08——裙楼
Phase 08—Podium

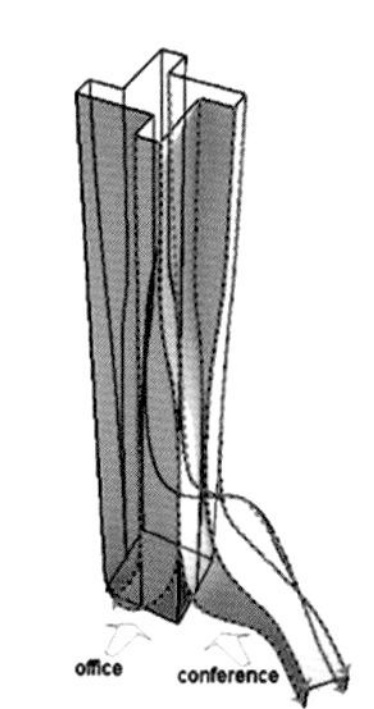

阶段 09——统一
Phase 09—Draping

4 5	6 7 8

4. 外立面细部
Facade details
5. 项目区位图
Location map
6. 陆上人视图与手绘图
External view and sketches
7. 分析图
Analysis diagram
8. 设计意向
Design concept

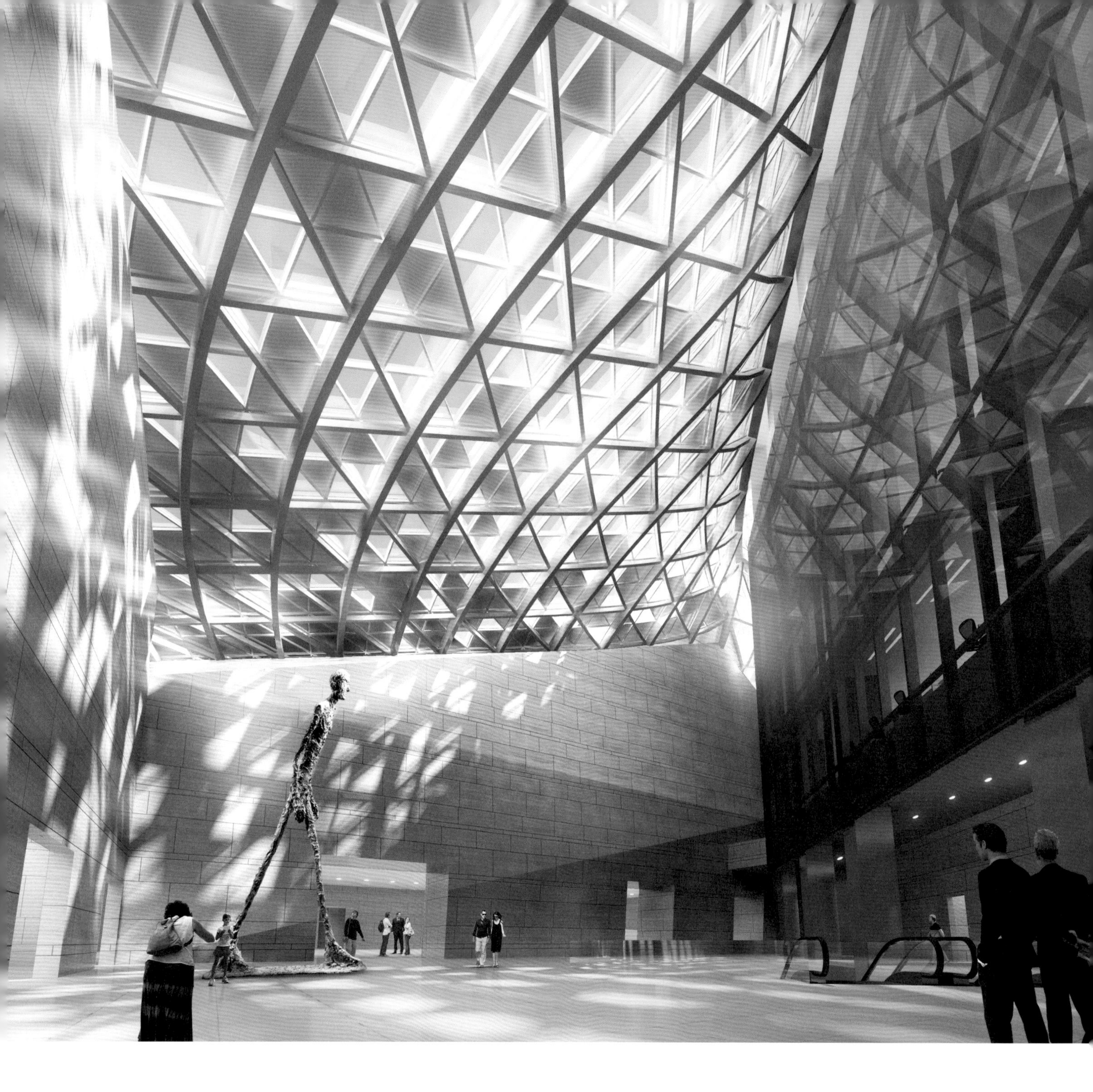

Location: A pearl in the Metropolis

Zhuhai is located on the west side of the Pearl River Delta, across the water from Shenzhen and Hong Kong and immediately adjacent to Macao. With the opening of the Hong Kong-Zhuhai-Macao Bridge, the three cities will be tightly integrated. Being the only city connecting Hong Kong and Macao by land and with only 30 minutes travel time between the three cities, the development potential of Zhuhai is likely to exceed that of Shenzhen. Hengqin International Financial Center is located on the island of Hengqin, inside Zhuhai's Shizimen Central Business District in the Xijiang River Delta, with Macao to the other side of the river. Shizimen Central Business District has been named the new financial hub in the overall Pearl River Delta plan. This building designed by Aedas will become the landmark of Hengqin Island.

Hengqin International Financial Center consists of a 333-metre-high office and residential tower overlooking the waterfront, and a podium containing commercial retail and convention facilities. From the exterior the building looks like four towers converging into one building, symbolising the convergence of talent from Zhuhai, Macao, Hong Kong and Shenzhen at Hengqin.

Inspiration: Chinese Dragon Culture

The dragon is the totem of Chinese civilisation. The design of the property was inspired by the famous painting "Nine Dragons" dating

9	10
	11

9. 内部中庭
Atrium
10. 三层平面图
Level 3 floor plan
11. 九层平面图
Level 9 floor plan

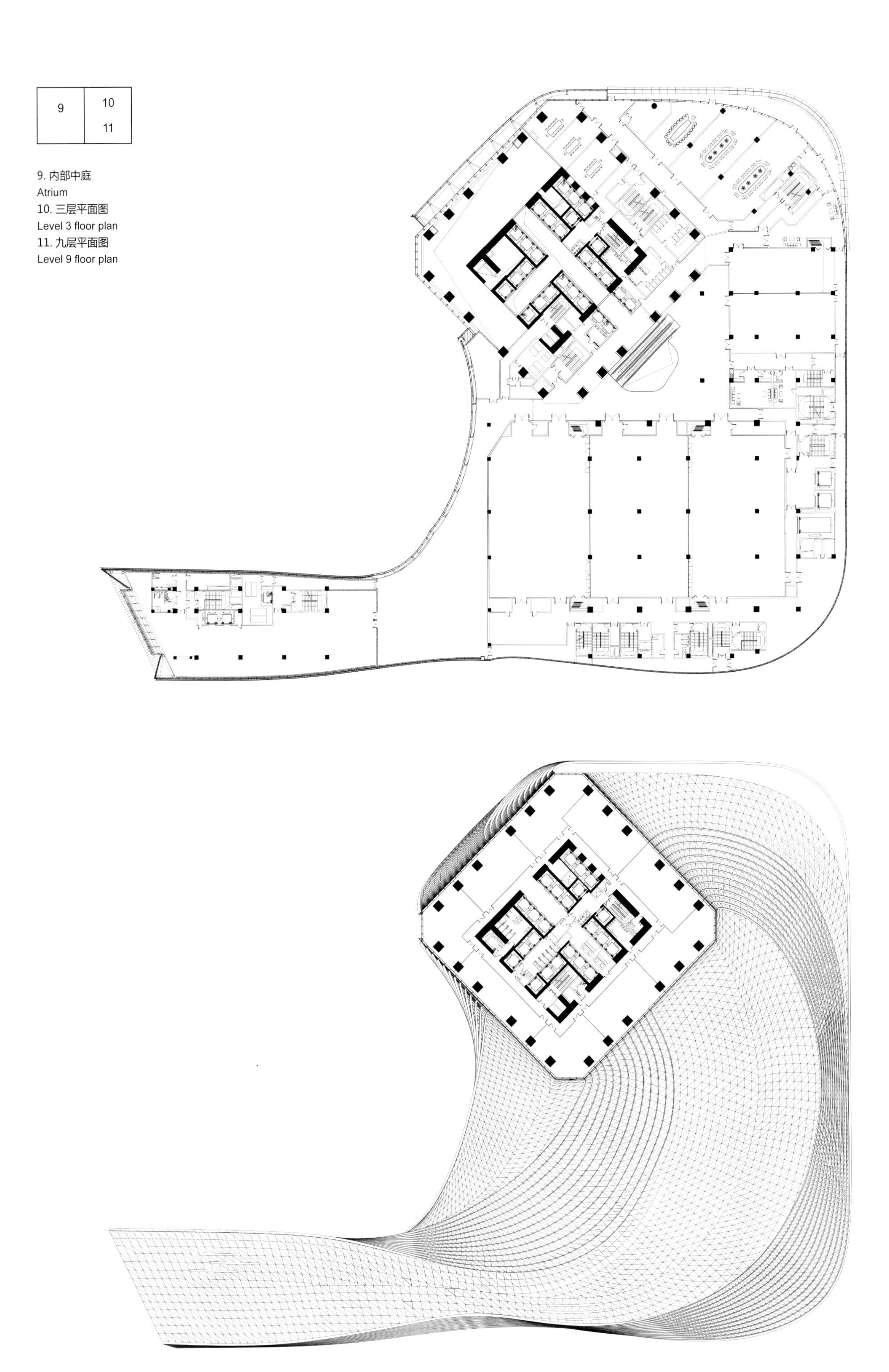

from the Southern Song Dynasty. In the painting, various forms of the holy dragons are crossing each other in stormy sea and changing clouds. Architects focused on the characteristics of the surrounding region from the very beginning of the design. The spirally rising volume evokes the well-known Chinese image of the Jiaolong — a water dragon — bursting forth from the sea. In ancient Chinese literature, the Jiaolong is a mythical creature capable of invoking storms and waves. It is often used to describe a newborn power.

Form: Streamlined Beauty

The facade of Hengqin International Financial Center is made of metal panels and glass curtain wall. The curtain drops down and covers the exhibition hall in the podium and becomes a canopy to shelter passers-by from the rain. Traditional tower-and-podium design is self-contained without enough linkages to the surrounding area. For this project, Aedas' architects drew, stretched and reversed the straight line between the tower and the podium, allowing the roof of the podium to be extended and raised. The exterior walls unite the soaring towers with the relatively wide and low podium.

Zhuhai is subjected to a long typhoon season every year, and the curvaceous shape of the Hengqin International Financial Center responded to this by reducing the wind load of the tower, making the shape lighter and more dynamic. To maximise great views, the floors change shape from square office floors to U-shaped serviced apartment floors. Moreover, a gigantic multi-function atrium is enclosed by the U-shaped apartment, presenting a beautiful scene for people living high in the air.

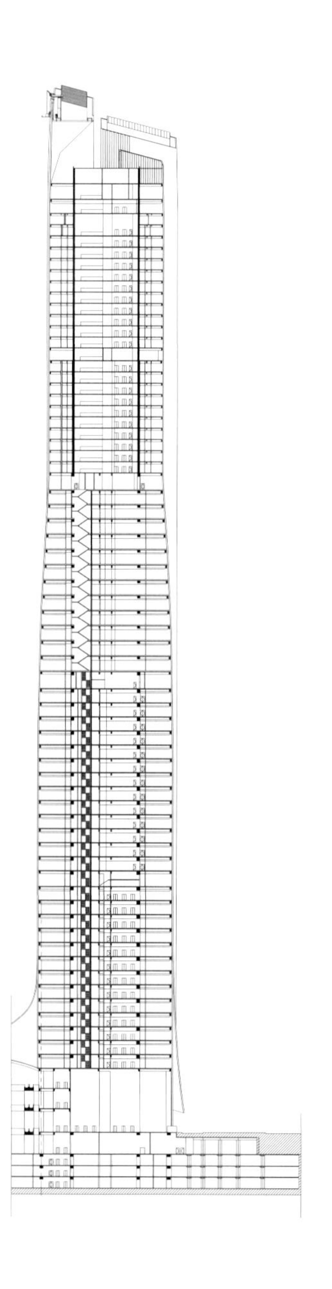

12 13	14

12.13. 剖面图
Section
14. 海上人视图
Perspective view from the sea

珠海横琴中冶总部大厦（二期）

Hengqin MCC Headquarters Complex (Phase II), Zhuhai

珠海横琴中冶总部大厦（二期）位于珠海市横琴新区港澳大道北侧，莲花大桥及横琴口岸的咽喉地段，屹立国门，俯瞰澳门。项目包含两栋塔楼、裙楼购物中心及室外商业街。购物中心塑造出表现力极强的商业空间，配合室外商街上一系列的景观节点营造出多层次文化旅游主题，不仅与塔楼动静辉映，更能有效地吸引客流，打造出令人难忘的一站式文化旅游目的地型的购物天堂。

以“双龙戏珠”为寓意，双塔楼造型展现出腾龙般的非凡气势和飘逸姿态，体现了丰富的华夏文化寓意。双塔“双龙”环抱“龙珠”，造型线条极具上升感；塔身立面设计取意“龙鳞”，形成丰富渐变的幕墙肌理。购物中心立意取自“莲花”，呼应莲花大桥这一独特地标，以形似莲叶的灵动表皮围合购物中心和商业街，以初绽莲花作为购物中心两个共享中庭的主题，以中心下沉广场为池心层层漾出阶梯和步道。双塔的设计以最大化海景视野为宗旨，平面布局高效，顶部设置观景平台。设计采用了大量室外连桥，并设置过街天桥与周边地块连接，在多个竖向高度层面引进人流、提高商业活力。购物中心以二层天桥强化出商业主要动线，又利用下沉广场激活地下商业，高效汇聚客流，完美地实现了商业价值的最大化。

1. 倚靠秀丽群山，面对浩瀚海洋，珠海是珠三角地区的一颗明珠
Cityscape of the coastal city Zhuhai
2. 塔楼外观
External view
3. 俯瞰塔楼及购物中心
Aerial view
4. 项目区位图
Location map

项目信息 | PROJECT INFORMATION

地　　点：中国珠海
建筑面积：486 257m²
设计时间：2015—2017 年
建成时间：2019 年
项目功能：综合体
业　　主：中冶国际投资发展有限公司
主要设计人：温子先博士，纪达夫

Location: Zhuhai, PRC
Gross floor area: 486,257 m^2
Design time: 2015-2017
Completion year: 2019
Sector: Mixed-use
Client: MCC International Investment Development Co., Ltd.
Directors: Dr. Andy WEN, Keith GRIFFITHS

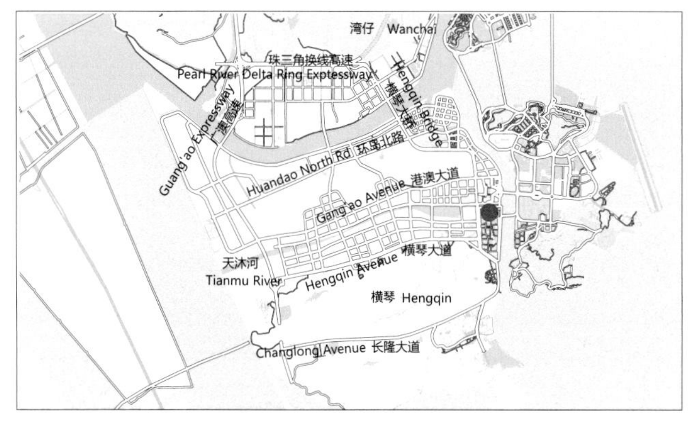

The Hengqin MCC Headquarters Complex (Phase II) sits in a prime area on Hengqin Island in Zhuhai, in close proximity to the Lotus Bridge that links the city to Macao's Cotai district. Overlooking some of the world's largest casino resorts across the river, the development is set to become an integrated culture and tourism-oriented complex, with two high-rise Grade-A office towers and a series of retail, banquet, entertainment and leisure venues. The highly expressive commercial space of the shopping mall, together with the outdoor shopping streets, creating an unforgettable one-stop destination for cultural tourism and a shopping paradise for visitors.

The design of the two towers and the small banquet building in between took reference from the mythological image of two dragons chasing a pearl. The towers portray a sense of strength and power with a facade that shimmers under the sun, while the banquet building looks like an exquisite pearl. The shopping center design, by contrast, echoes the nearby Lotus Bridge. It resembles a blooming lotus flower with "lotus leaves" wrapping around the shopping center and lining up the retail streets. The design of the towers is aimed at maximising the sea view; the towers have highly efficient layout plan and a viewing platform on the top. The design includes outdoor bridges that connect the complex and the neighbouring sites at multiple levels, attracting pedestrian flow to the retail area. The shopping center has a double layer of bridges to enhance circulation and a sunken plaza to activate the underground retail, which brings together different pedestrian flows, maximising commercial profit.

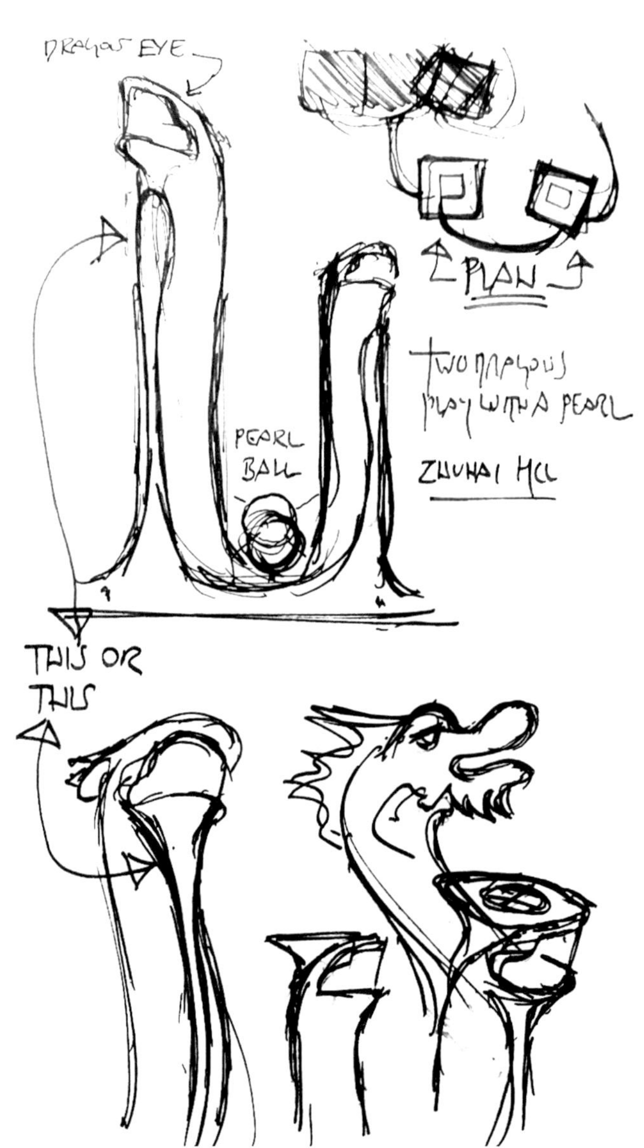

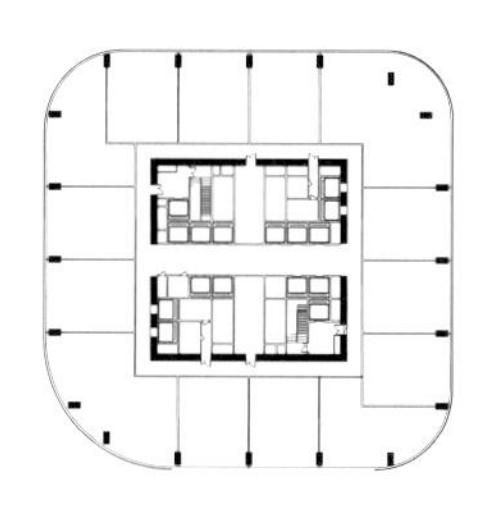

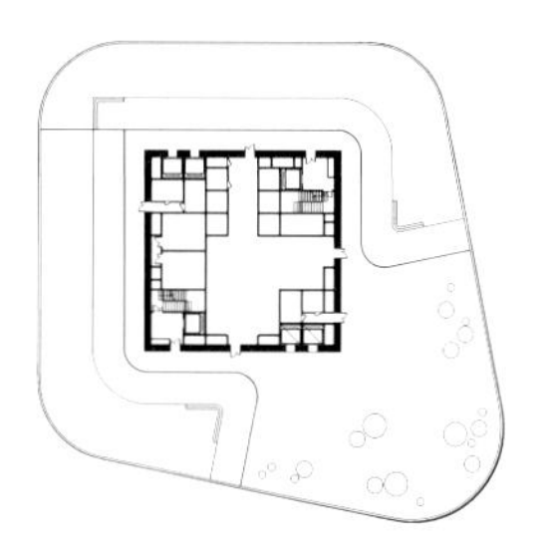

5	7 8 9
6	10

5. 购物商街
Retail street
6. 手绘图
Sketches
7. 一层平面图
Level 1 floor plan
8. 塔楼办公低区标准层平面
Typical office floor plan
9. 塔楼空中观景层平面图
Floor plan of observation deck
10. 下沉广场
Sunken plaza

中国香港地区和中国台湾地区
HONG KONG AND TAIWAN REGIONS OF CHINA

中国大陆长江以北 NORTH OF YANGTZE RIVER

2 3 4 7 大连恒隆广场
Olympia 66, Dalian

2 4 6 北京大望京综合开发项目
Da Wang Jing Mixed-use Development, Beijing

3 4 北京大兴 3 及 4 地块项目
Daxing Plots 3 and 4, Beijing

1 3 6 北京新浪总部大楼
Sina Plaza, Beijing

1 2 3 4 6 北京北苑北辰综合体
North Star Mixed-use Development, Beijing

2 3 4 7 青岛金茂湾购物中心
Jinmao Harbour Shopping Center, Qingdao

1 2 4 成都恒大广场
Evergrande Plaza, Chengdu

1 7 重庆新华书店集团公司解放碑时尚文化城
Xinhua Bookstore Group Jiefangbei Book City Mixed-use Project, Chongqing

2 3 4 7 武汉恒隆广场
Heartland 66, Wuhan

中国大陆长江以南 SOUTH OF YANGTZE RIVER

3 6 7 上海星荟中心
Shanghai Landmark Center, Shanghai

1 2 3 4 6 上海龙湖虹桥项目
Longfor Hongqiao Mixed-use Project, Shanghai

6 7 上海虹桥世界中心
Hongqiao World Center, Shanghai

2 3 4 6 7 无锡恒隆广场
Center 66, Wuxi

3 7 苏州西交利物浦大学中心楼
Xi'an Jiaotong-Liverpool University Central Building, Suzhou

2 3 义乌之心
The Heart of Yiwu, Yiwu

2 3 6 广州南丰商业、酒店及展览综合大楼
Nanfung Commercial, Hospitality and Exhibition Complex, Guangzhou

3 8 广州邦华环球贸易中心
Bravo PARK PLACE, Guangzhou

3 珠海粤澳合作中医药科技产业园总部大楼
Headquarters, Traditional Chinese Medicine Science and Technology Industrial Park of Co-operation between Guangdong and Macao, Zhuhai

3 8 深圳宝安国际机场卫星厅
Shenzhen Airport Satellite Concourse, Shenzhen

1 7 珠海横琴国际金融中心
Hengqin International Financial Center, Zhuhai

2 3 4 珠海横琴中冶总部大厦（二期）
Hengqin MCC Headquarters Complex (Phase II), Zhuhai

中国香港、台湾地区都曾经被誉为 20 世纪七八十年代的“亚洲四小龙”，是亚洲经济快速发展的地区。城市建设与建筑繁荣为亚洲高密度人居环境建设留下了经验与教训，21 世纪的全球化和中国大陆的发展进步给港台地区以紧迫感。Aedas 在港台的交通建筑、学校建筑等方面的创新给城市带来新的景象。

Hong Kong and Taiwan were two of the “Four Asian Dragons” in the 1970s and 1980s and enjoyed rapid economic development in Asia. The high-density urban development has left us with experience and lessons. Globalisation in the 21st century and development of mainland China are putting pressure on Hong Kong and Taiwan. Innovation in transportation facilities, schools and other buildings designed by Aedas for Hong Kong and Taiwan brings new cityscape to the two regions.

Ⓒ 中国香港地区和中国台湾地区 HONG KONG AND TAIWAN REGIONS OF CHINA

- ❶❺❽ 香港西九龙站
  Hong Kong West Kowloon Station, Hong Kong
- ❷❸❺❻ 港珠澳大桥香港口岸旅检大楼
  Hong Kong-Zhuhai-Macao Bridge Hong Kong Port - Passenger Clearance Building, Hong Kong
- ❶❺❽ 香港国际机场中场客运廊
  Hong Kong International Airport Midfield Concourse, Hong Kong
- ❶❺❽ 香港国际机场北卫星客运廊
  Hong Kong International Airport North Satellite Concourse, Hong Kong
- ❹ 香港富临阁
  The Forum, Hong Kong
- ❸❼ 台北砳建筑
  Lè Architecture, Taipei
- ❸ 台中商业银行企业总部综合项目
  Commercial Bank Headquarters Mixed-use Project, Taichung

Ⓓ 一带一路 BELT AND ROAD

- ❸❹❽ 新加坡星宇项目
  The Star, Singapore
- ❸❽ 新加坡 Sandcrawler
  Sandcrawler, Singapore
- ❽ 阿联酋迪拜 Ocean Heights
  Ocean Heights, Dubai, UAE
- ❺❽ 阿联酋迪拜地铁站
  Dubai Metro, Dubai, UAE
- ❽❾ 英国唐卡斯特 Cast 剧院
  Cast, Doncaster, UK

九项设计理念 NINE POINTS OF DESIGN IDEAS

1. 《国家新型城镇化规划（2014—2020 年）》
   NATIONAL NEW URBANISATION PLAN (2014-2020)
2. 高密度城市枢纽
   HIGH DENSITY CITY HUBS
3. 具有通透性和连接性的设计
   POROUS AND CONNECTED DESIGNS
4. 新型商业零售
   THE NEW RETAIL
5. 基础设施设计
   INFRASTRUCTURE DESIGN
6. 中国的特大城市与城市设计
   CHINA MEGAPOLIS AND URBAN DESIGN
7. 文化和故事
   CULTURE AND STORY
8. 一带一路
   BELT AND ROAD
9. 城市改造与修复
   URBAN RENEWAL AND RESTORATION

香港西九龙站

Hong Kong West Kowloon Station, Hong Kong

打造“港粤一小时生活圈”

香港位于中国南海沿岸，向西是澳门和广东珠海，北面是广东深圳，南面是万山群岛。它既是一座高度繁荣的国际大都市，也是中西方文化交融之地，素有“东方之珠”“美食天堂”和“购物天堂”等美誉，也是全球最富裕、经济最发达和生活水准最高的地区之一。香港西九龙站立于香港的门槛，承担着重要的联系职能，而西九龙铁路的特殊意义在于：它是向北通往北京的新建铁路的门户，将构成香港历史上最大型的铁路运输网络。西九龙铁路总站的运作模式更接近于一个机场，在同一座设施内同时设有香港和内地的入境处，通过既有铁路与中环心脏地带紧密相连。配合九龙站新的商业发展，西九龙站将成为重要的商业中心及铁路枢纽。

全球最大的地下高速铁路车站

作为连接香港和内地首条高速铁路在香港段的终点站，车站设有 15 条铁路轨道，仅可用面积就超过 400 000m^2。整体分为 4 个主要地下楼层：其中地下一层为售票

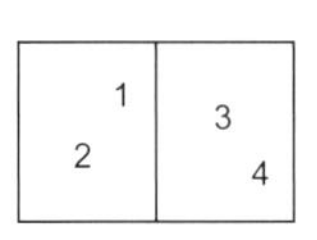

1. 香港城市景观
Hong Kong cityscape
2. 建筑夜景
Night view
3. 屋顶带型景观走道
Rooftop walkway
4. 项目区位图
Location map

Photography © Virgile Bertrand

项目信息 | PROJECT INFORMATION

地　　点：中国香港
可用面积：超过 400 000m²
设计时间：2009—2015 年
建成时间：2018 年
项目功能：基础设施
业　　主：香港铁路有限公司
主要设计人：Andrew BROMBERG

Location: Hong Kong, PRC
Usable floor area: Over 400,000 m²
Design time: 2009-2015
Completion year: 2018
Sector: Infrastructure
Client: MTR Corporation Ltd.
Design Director: Andrew BROMBERG

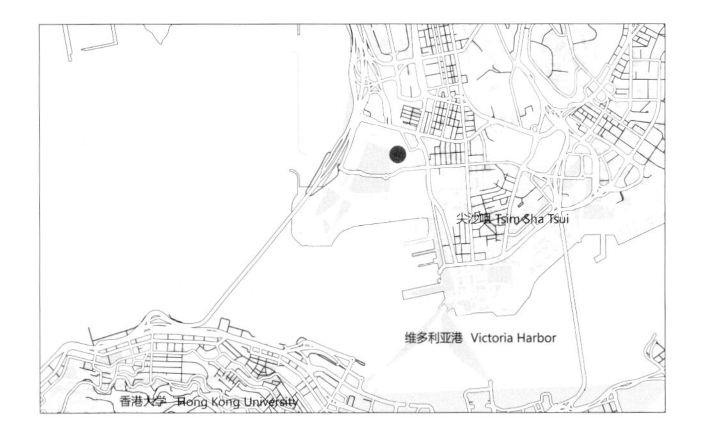

Photography © Virgile Bertrand

大厅，地下二层为抵港层，地下三层为离港层，地下四层为站台。车站设计为场地引入超过 3hm^2 的“绿色广场”，连通高铁总站地面楼层、九龙站、柯士甸站及西九龙文化区，为市民和旅客提供优质的公共绿化休憩空间及舒适的步行环境。西九龙站是全球最大的地下高速铁路车站之一。作为香港的“门户”，车站不但与周边的城市环境联系与互动，还让抵港或离港的旅客切身体会到“身处香港”的在地感。

探索：营造“原始森林”的火车站

出发乘客需要宽敞的候车空间，整个团队经过深入的研究将所有结构支持构件有效地集中起来，为地下离境大堂的巨型中庭腾出宽裕的场地。户外地面层向下弯曲，屋顶结构朝向海港，形成约 45m 高的场地。旅客会将注意力集中在南立面，透过它可远眺中环天际线、太平山顶等风景。透过轨道层和站台层中间的坑道向下俯瞰，又别有一番景象。大多数传统的火车站就像巨大的工业洞穴，西九龙站则不同。车站天幕呈流线型，由 4 000 多块大型玻璃组件以及 16 000 多块铝板构成，其中 90% 为不规则形状，垂直高度达 50m，施工难度极高。作为主要力学支撑的 V 型构架，像一棵棵参天大树张开枝桠，让人产生置身原始森林的联想。

从建筑外部看，巨型屋面呈环抱之势，加以层层叠叠的屋顶花园及观景台，展现出自然友好的城市界面。南部通道通往规划中的西九龙文化区大型公共广场。车站西部是用于室外表演的圆形剧场，剧场通过一系列金属刀片造型构件（即出发大厅上巨大的叶片的室外部分）与屋面上种满植物的雕塑花园相连。

Creating One-hour Living Circle of Hong Kong-Guangdong

Hong Kong is located at the coast of the South China Sea, east of Macao and Zhuhai, south of Shenzhen and north of the Wanshan Islands. Hong Kong is a prosperous international metropolis and a place where Chinese and Western cultures blend together. It is known as the “Pearl of the Orient” and a paradise for food and shopping. It is also one of the world's richest and most economically developed regions, with a high standard of living. Hong Kong West Kowloon Station is a new gateway to the city and it plays an important role in connecting Hong Kong and mainland China. Aedas has decades of experience in designing railway transportation facilities in Southeast Asia. The essential significance of Hong Kong West Kowloon Station is two-fold: it will be the gateway of Hong Kong on the new railway that eventually stretches up to Beijing; and it is going to be the largest railway transportation hub in Hong Kong’s history. The station will function more like an airport. In particular, the station will accommodate the immigration checkpoints of both Hong Kong and mainland China. It will be connected to Central, Hong Kong’s central business district, through existing underground railways. Hong Kong West Kowloon Station anchors a new development area that will make West Kowloon an important business center and transit hub.

One of the Largest Underground Railway Stations in the World

As the terminus for the first high-speed railway connecting Hong Kong to the mainland, the usable floor area is over 400,000

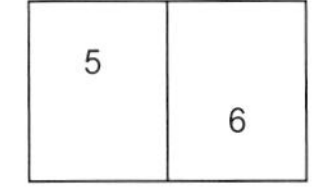

5. 户外广场
Playspace
6. 分析图
Analysis diagram

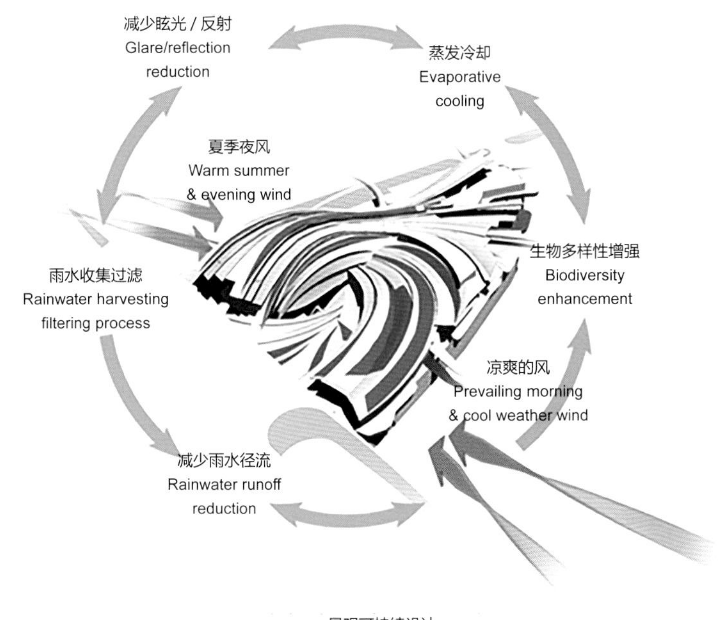

景观可持续设计
Sustainable landscape design

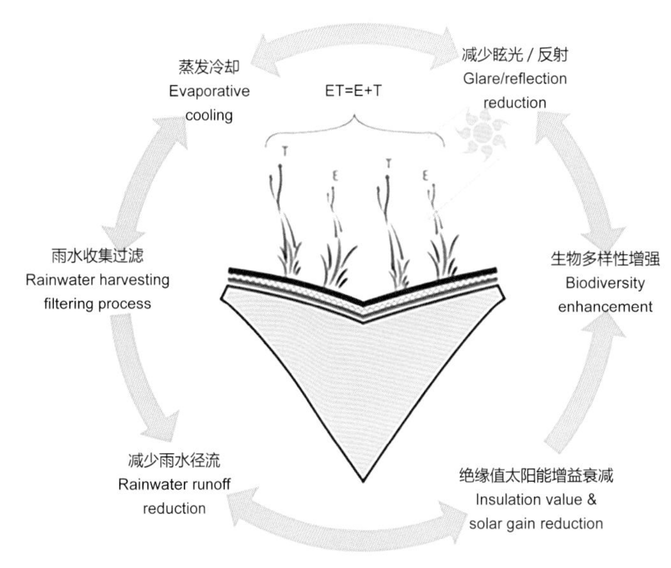

屋顶结构可持续设计
Roof structure sustainable design integration

Photography © Virgile Bertrand

square metres, with 15 railway tracks. The station comprises four underground floors and a main entrance hall on the ground; B1 is the concourse; B2 is the arrival hall; B3 is the departure hall; and B4 is where the platforms are located. The design of the station introduces over three hectares of "green plaza" to the site, reaching out to the nearby Kowloon MTR Station, Austin MTR Station and the future West Kowloon Cultural District, providing locals and tourists high quality public space and a comfortable walking environment. As one of the largest underground railway stations in the world and the gateway to Hong Kong, it is considered vital to connect the station with the surrounding urban context and make arriving or departing passengers aware that they are in Hong Kong.

Exploration: Train Station as an Old-growth Forest

As departing passengers need a spacious waiting area, the design team conducted an in-depth study that led them to concentrate all of the station's supporting structure components. This allowed for a very large void that penetrates down into the departure hall below. The outside ground plane is slanted down towards the hall, and the roof gestures out towards the harbour. The result is a 45-meter high volume that focuses all attention through the south facade towards views of the Central skyline and Victoria Peak beyond. There are also intriguing glimpses through gaps in the floor of the rail tracks and platforms far below. Most Chinese railway stations are vast industrial caverns, but not Hong Kong West Kowloon Station. The enormous roof is streamlined and consists of more than 4,000

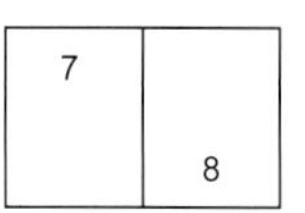

7. 屋顶带型景观走道
Rooftop walkway
8. 西九龙站屋顶起伏有致，使候车空间更丰富
The undulating roof form enriches the space

large-scale glass components and over 16,000 pieces of aluminium panels, and 90 percent of which are irregular in shape and increase the difficulty in construction. The vertical height is approximately 50 meters. The V-shape structures bring to mind towering trees, creating an atmosphere like an old-growth forest.

From outside, the enormous roof appears to hug the ground. With layered, landscaped roof terraces and observation deck, it creates a natural kind of cityscape. The southern path leads to the future West Kowloon Cultural District. On the west is an outdoor performance amphitheatre that is linked to the roofscape and the heavily planted sculpture garden. Hong Kong West Kowloon Station is one of the largest underground railway stations in the world.

Photography © Virgile Bertrand

Photography © Virgile Bertrand

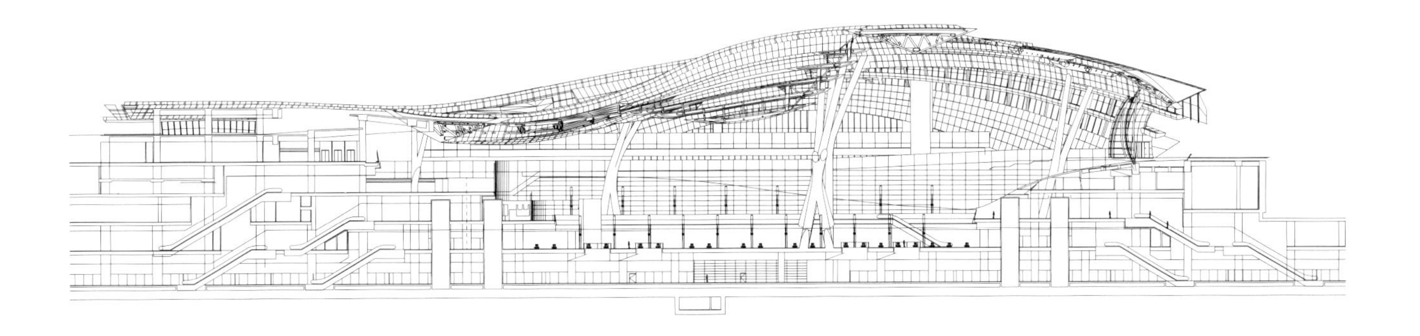

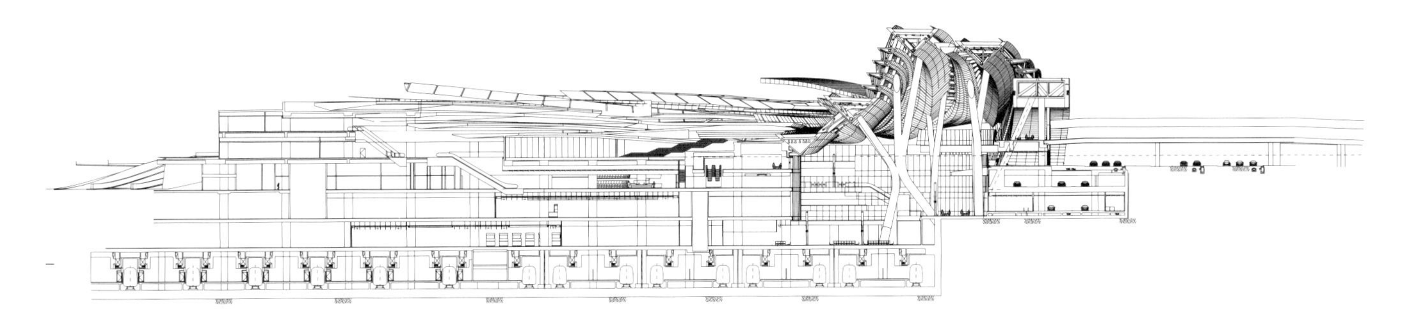

9	10 11 12

9. 出发大厅室内
Interior of departure levels
10. 11. 剖面图
Section
12. 西九龙站屋顶起伏有致，使候车空间更丰富
The undulating roof form enriches the space

港珠澳大桥香港口岸旅检大楼

Hong Kong-Zhuhai-Macao Bridge Hong Kong Port - Passenger Clearance Building, Hong Kong

与罗杰斯史达克哈伯建筑事务所 (RSHP) 联合设计

Designed in joint venture with Rogers Stirk Harbour + Partners (RSHP)

香港、珠海、澳门三地隔海相望，为三座极具代表性的港口城市：他们同时保有着港口城市开放包容的特性，又各自拥有不同的城市文化与历史。港珠澳大桥是在“一国两制”的条件下，由粤港澳三地首次合作共建的超大型基础设施项目，其中的香港口岸旅检大楼，为香港提供了全新的抵境门户。

港珠澳大桥香港口岸旅检大楼位于香港国际机场东北面海域中一座 $150hm^2$ 的新造人工岛之上，由 Aedas 与 RSHP 共同打造，与相邻的海天客运码头、港铁机场快线及东涌线共同组成香港交通枢纽。整座建筑时刻处于“动态”之中，承载着巴士、轿车、货车与行人的通行。设计者对场地内部及周边动线进行了细致考虑，通过波浪状屋顶的不同色彩来强化设施的功能分区，使道路指向更加清晰可辨。设计在建筑内通行路线中穿插通高的峡谷状中庭，在引入自然光线至建筑各层的同时，确保了与线型屋顶的视觉连接。建筑整体形态优雅简洁，不仅与香港气候和环境呼应，还通过模块化的场外预制加工，保证了高效的施工过程并呈现出最高标准的建筑品质。

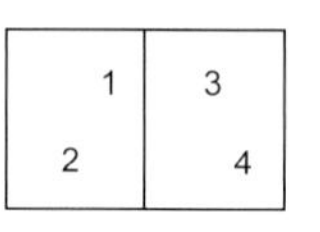

1. 香港青马大桥
Hong Kong Tsing Ma Bridge
2. 港珠澳大桥香港口岸建在人工岛上
The Hong Kong-Zhuhai-Macao Bridge Hong Kong Port is constructed on an artificial island
3. 港珠澳大桥香港口岸旅检大楼室内
Interior of the Passenger Clearance Building
4. 项目区位图
Location map

项目信息 | PROJECT INFORMATION

地　　点：中国香港
建筑面积：超过 90 000m^2
设计时间：2010—2018 年
建成时间：2018 年
项目功能：基础设施
主要设计人：纪达夫，江立文
　　理查德·保罗 (RSHP)

Location: Hong Kong, PRC
Total floor area: Over 90,000 m^2
Design time: 2010-2018
Completion year: 2018
Sector: Infrastructure
Directors: Keith GRIFFITHS, Max CONNOP
　　Richard PAUL (RSHP)

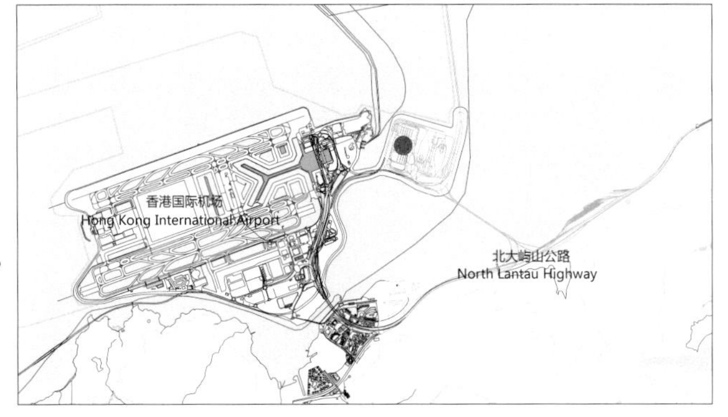

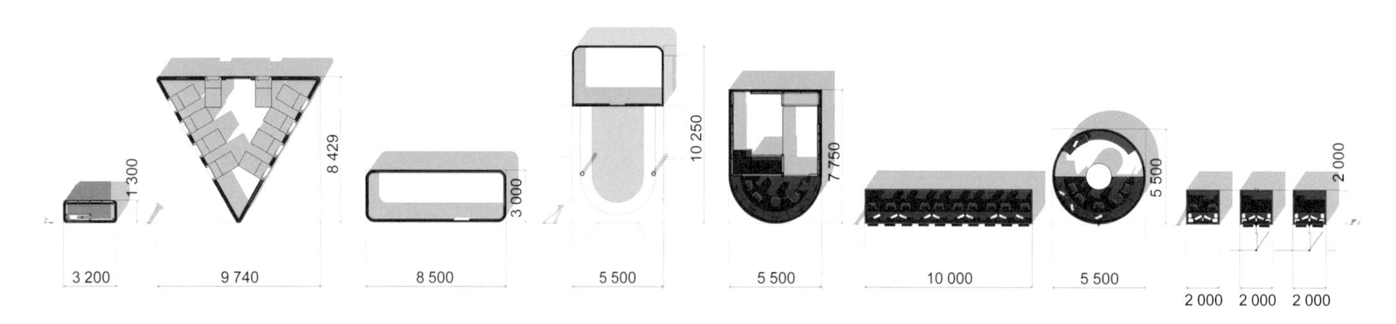
1 300
8 429
3 000
10 250
7 750
5 500
2 000
3 200
9 740
8 500
5 500
5 500
10 000
5 500
2 000
2 000
2 000

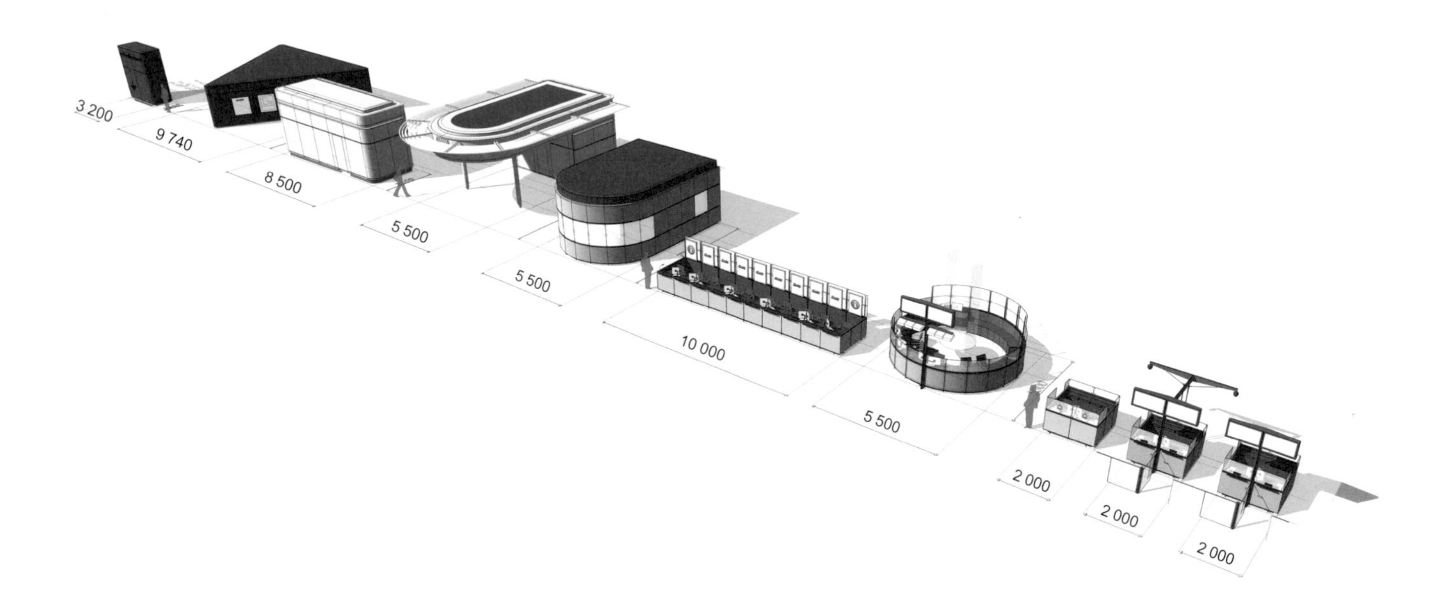
3 200
9 740
8 500
5 500
5 500
10 000
5 500
2 000
2 000
2 000

5	8
6	
7	9

5. 波浪状的屋顶
The waving form roof
6. 7. 各项设施的平面和轴测图
Plans and axonometric diagrams of various facilities
8. 旅检大厅
Immigration hall
9. 屋顶设计延伸到室外
The roof extends to the outside of the building

Hong Kong, Zhuhai and Macao are three port cities with the same spirit of openness and inclusivity, but with markedly different cultures and histories. The Hong Kong-Zhuhai-Macao Bridge is the first piece of mega-infrastructure built collaboratively by the three cities under the "One Country, Two Systems" policy.

The bridge requires a new customs and immigration facility that serves as a new entry point to Hong Kong. Located on a man-made 150-hectare island north of the Hong Kong International Airport, the Passenger Clearance Building (PCB) is jointly designed by Aedas and RSHP. Together with the SkyPier, MTR Airport Express and Tung Chung Line, they form a transport hub connecting Hong Kong to mainland China and the rest of the world.

The PCB will be constantly filled with movements: buses arriving and leaving, and cars and lorries waiting to be processed. Careful thought has therefore been put into how users will move around on the site. The simple, clear circulation through the facility is reinforced by the different color-codings of the waving roof, which enhance legibility and provides wayfinding. The movement through the building is punctuated by full height canyon-shaped atriums allowing the penetration of natural daylight to all levels of the building, ensuring there is a visual connection to the linear roof form to further reinforce clarity of wayfinding.

The building is simple and elegant. While the design adapts to the climate and environment of Hong Kong, offsite prefabrication of the modular roof has enabled an efficient construction process, achieving a very high level of quality.

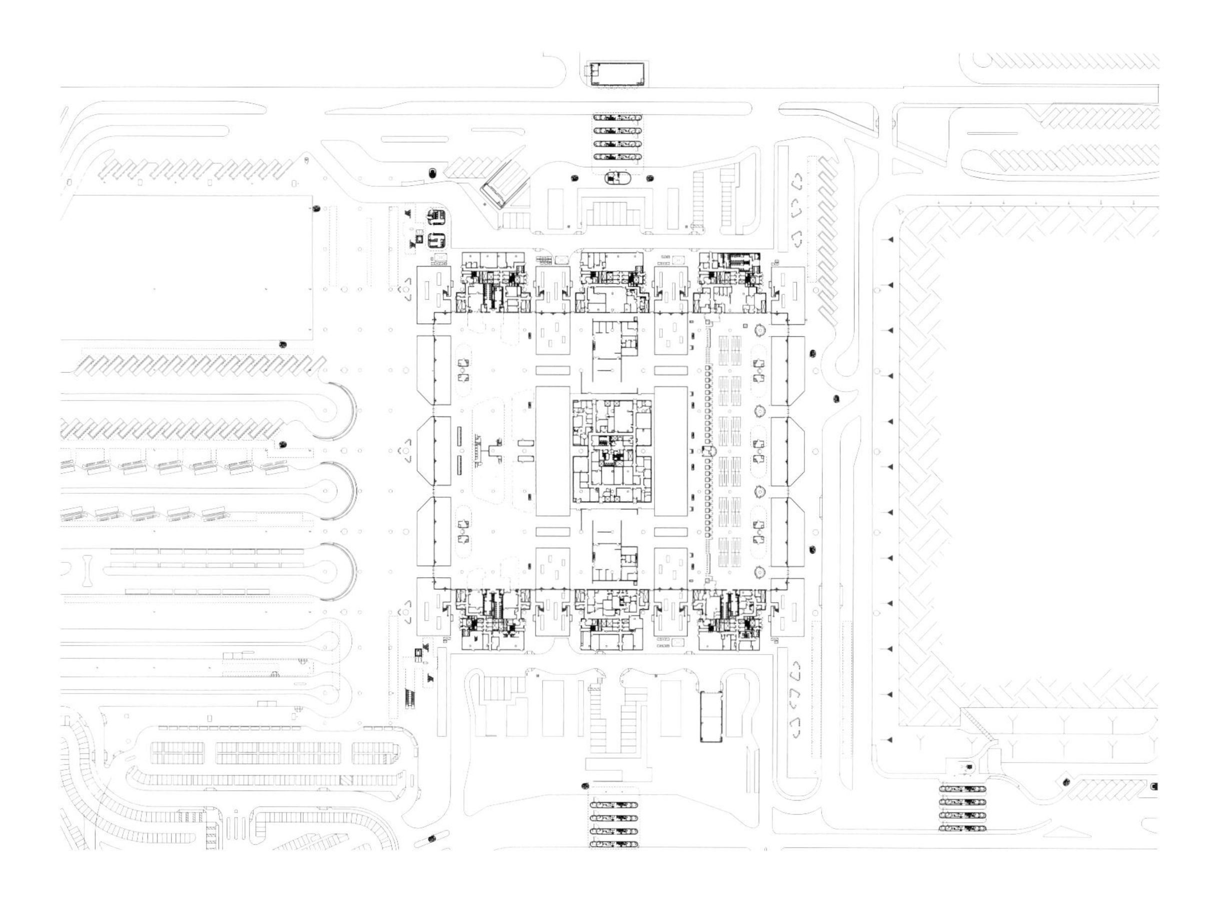

10	11 12

10. 旅检大厅
Immigration hall
11. 地下一层平面图
Ground floor plan
12. 一层平面图
Level 1 floor plan

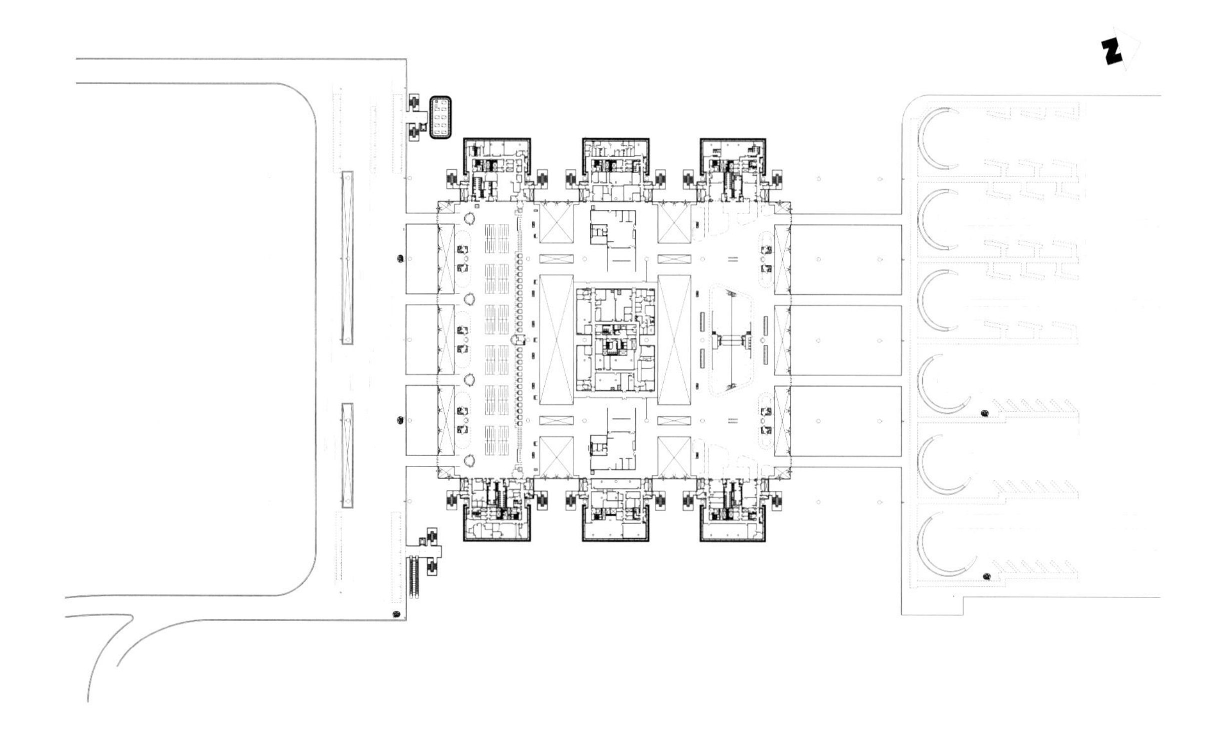

HONG
KONG

13	14
	15

13.14. 中庭
Atrium
15. 剖面图
Section

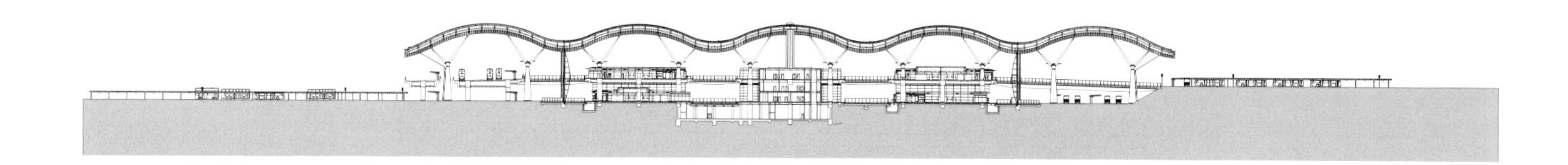

香港国际机场中场客运廊

Hong Kong International Airport Midfield Concourse, Hong Kong

在北卫星客运廊成功建造并投入运营的 5 年之后，Aedas 又为香港国际机场新添了一座客运廊——中场客运廊。它位于一号航站楼的西侧，坐落于两条既有跑道之间，共 5 层，建筑总面积达 10.5 万 m^2，设有 20 个停机位，每年可额外处理约 1 000 万人次的客运量。

造型：西晒的巧妙规避

项目采用“建筑呼应环境”的模式，根据建筑的朝向来确定屋顶形状和玻璃幕墙的最佳方案。香港西晒严重，过热的温度和强烈的光线会大大降低候机人群的舒适度。Aedas 采用了不对称的屋面设计，沿南北轴修建，将西面的屋面延伸、突出以减少西侧太阳辐射。团队更匠心独具地为大厅设计了朝向北面倾斜的天窗，可以在较小太阳辐射条件下补充采光。

能源：绿色节能的设计

中场客运廊的建设无疑为香港机场增加了客容量，由此建筑设计大量采用了可持续理念，包括废物废水循环、自然冷却、再生水冷却等。为了最大化客运廊的效率和节能特性，团队采用了逾 35 项绿色设计，如在采光照明方面采用了节能照明系统、高性能玻璃和遮阳朝北的天窗，在最大限度地利用自然光照的同时，降低太阳辐射热和空调用电。与此同时，还有在空调系统中使用循环水制冷、以海水冲厕等设计举措。另外，大楼楼顶安装了逾 1 200m^2 的光伏太阳能电池板，以收集可再生能源。

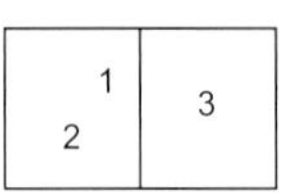

1. 香港城市景观
Cityscape of Hong Kong
2. 跑道之间的中场客运廊
Midfield Concourse located between two runways
3. 候机大厅室内
Interior of departure level

209
209
210
207

项目信息 | PROJECT INFORMATION

地　　点：中国香港	Location: Hong Kong, PRC
建筑面积：105 000m^2	Gross floor area: 105,000 m^2
设计时间：2010—2011 年	Design time: 2010-2011
建成时间：2015 年	Completion year: 2015
项目功能：基础设施	Sector: Infrastructure
业　　主：香港机场管理局	Client: Airport Authority Hong Kong
主要设计人：江立文，唐宙行	Directors: Max CONNOP, Albert TONG

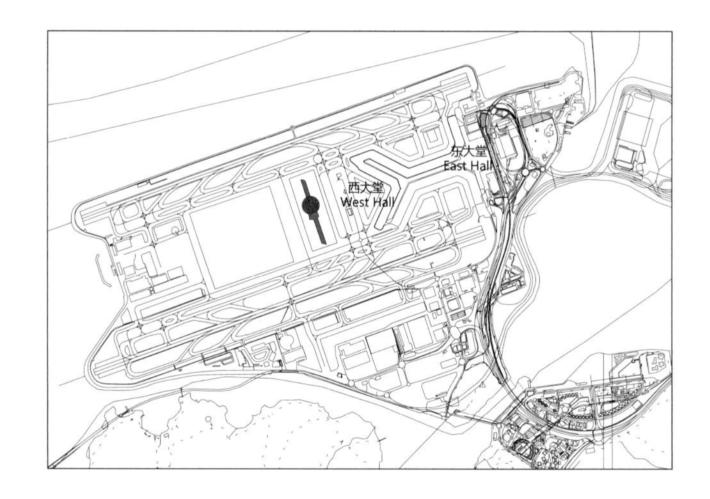

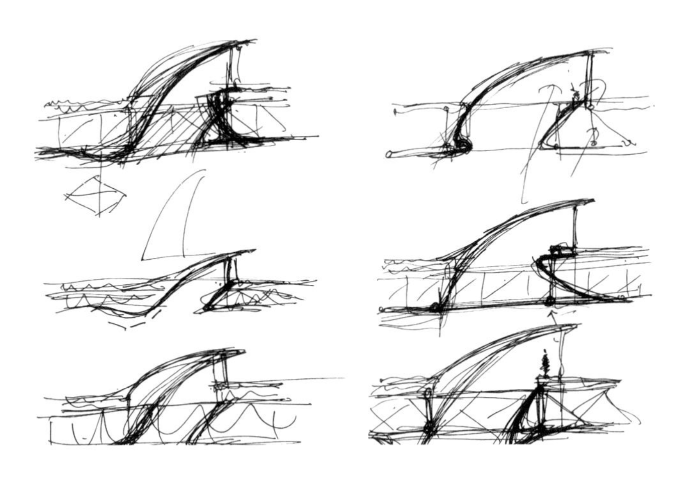

4. 出发大厅室内
Interior of departure levels
5. 项目区位图
Location map
6. 手绘图
Sketches
7.8. 自动捷运系统大厅
Automated people mover concourse

装饰：在地性的艺术

如何能让旅客在到达机场的第一瞬间就获得归属感？香港国际机场的中场客运廊给出了完美的解决方案：在自动捷运系统大厅有两幅巨幕玻璃抽象艺术壁画，由艺术家 Graham Jones 创作。入境处的红色是香港的标识，寓意着热情、活力、节日、激动，大块欢脱跳跃的色彩映照在抵港旅人的眼中，可以瞬间解除他们在旅行中的疲惫，加入城市的脉动中；出境处的蓝色是远行的象征，寓意着天空、安宁、探索、憧憬，其中点染的紫色，更为旅行添加了对神秘与未知的好奇。

地毯图案来自于维多利亚港夜景。设计师分解了照片，用抽象的方法提取那绚烂的颜色和水中倒影的元素，风格化地拼接出中场航站楼地毯的图案，这也是一款“高定”设计。香港的机场候机厅，以一系列情景元素组合而成的空间装饰，唤起人们潜意识中的认同感。

Aedas 作为主建筑设计师，同主顾问和工程师莫特麦克唐纳与奥雅纳组建的联合体，以及航空规划师 OTC 共同完成此项目。

Five years after the successful construction and operation of the North Satellite Concourse, Aedas has completed a new concourse — the Hong Kong International Airport (HKIA) Midfield Concourse. The Midfield Concourse is located to the west of Terminal 1 and between the two existing runways of HKIA. The project is a five-level concourse with a total floor area of 105,000 square meters. There are 20 aircraft parking stands. It can handle an additional 10 million passengers a year.

Form: Avoid Direct Sunshine from the West

The concourse adopts an environmentally responsive building form, with its roof shape and glazed facades optimised based on the building's orientation. Aedas designed an asymmetrical roof along the north-south axis, with the west eave extending out to reduce solar radiation. The project team ingeniously designed skylights that are rotated to face north to direct necessary natural daylight with less

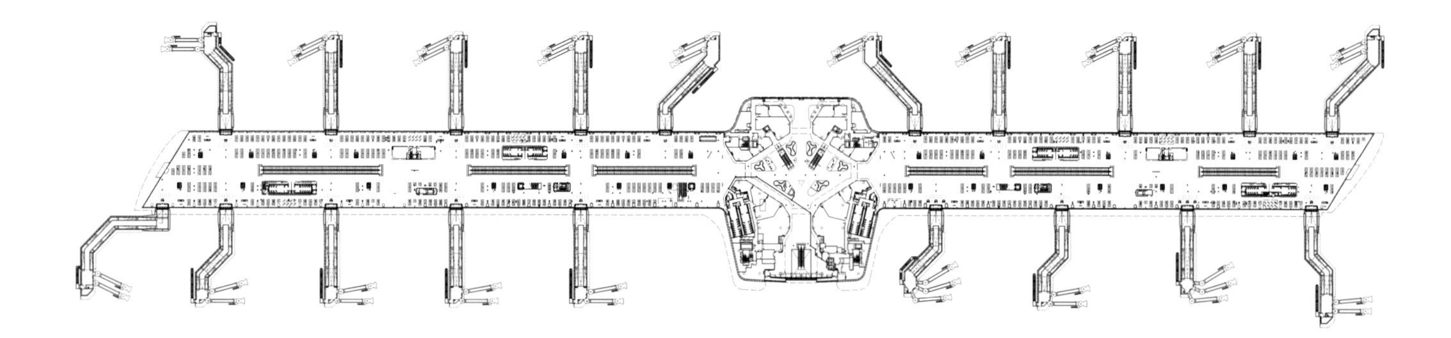

9. 一层平面图
Level 1 floor plan
10. 候机厅
Departures hall

solar radiation to the concourse.

Energy: Green, Energy-saving Design

The Midfield Concourse has undoubtedly increased the passenger capacity of HKIA. To maximise efficiency and sustainability, the design team has introduced more than 35 green features, including low-energy lighting systems, high performance glazing panels, solar shading and north-facing skylights to maximise natural lighting while reducing solar heat gain and saving on air-conditioning. Recycled water is used for the water-cooled chillers of the air-conditioning system and seawater for flushing. More than 1,200 square meters of solar panels were installed on the roof to harness renewable energy.

Decoration: Local Art

How can passengers get a sense of belonging when they first arrive at the airport? HKIA Midfield Concourse provides the perfect solution. There are two impressive cast glass walls by artist Graham Jones at the automated people mover concourse. On arrival, there is a red one; the colour is symbolising passion, vitality, festival and excitement, which is refreshing for passengers and an introduction to the dynamic city life. The blue wall in departures symbolises sky, peace, exploration, expectation, which, combined with a purple touch, highlights the mysterious and unpredictable adventure.

The design of the carpet stems from a night time picture of Victoria Harbour, with the vibrant colours and reflections in the water as the key features. The designers pixelated the picture and extracted elements to compose the image of the carpet, which was tailor made for this project. The local elements used for interior decoration are intended to make passengers aware that they are indeed in Hong Kong.

Aedas is the Lead Design Architect, working with Lead Consultant and Engineers Mott Macdonald and Arup Joint Venture, and OTC as Aviation Planners.

香港国际机场北卫星客运廊

Hong Kong International Airport North Satellite Concourse, Hong Kong

香港国际机场位于中国香港特别行政区的新界大屿山赤鱲角，自启用以来，先后逾 60 次获选为全球最佳机场，2016 年其旅客吞吐量更达到了 7 000 万人次。2009 年，北卫星客运廊建成，它是机场中一座独立运作的新卫星客运大楼，拥有 10 个附设登机桥的停机位，以接驳车辆连接一号客运大楼。北卫星客运廊以其充满现代感的造型、独特的空间处理，成为香港机场中一道亮丽的风景线。

造型：均好性的权衡

北卫星客运廊的屋顶如同一段行云流水般的海浪线，节制地表达出韵律感。精彩的形态并非单一出于对造型设计的思考，而是综合衡量了场地的复杂限制、设备系统的整体安置、结构设计的合理性及内部功能空间的使用需求等多方面因素后推导得出的解决方案。由于总平面布局紧凑，停机坪与北卫星客运廊的地面空间非常有限，如果将机电系统设施放置于下部很可能造成建筑屋顶高于周边建筑限高。Aedas 创造性地将所有大型机电系统与通风、照明设备等都放在建筑顶部。不同于

1 2	3

1. 作为城市重要的基础设施，香港国际机场点亮了城市文化
Not only an important piece of infrastructure, Hong Kong International Airport is also an emblem of the city
2. 傍晚的北卫星客运廊外景
External view
3. 北卫星客运廊屋顶起伏有致，使候机空间更丰富
The wavy roof offers passengers a rich spatial experience

DUTY FREE
DFS
NOBLETIME
BONUS GIFT
GET A BAG
580

项目信息 ｜ PROJECT INFORMATION

地　　点：中国香港
建筑面积：18 900m^2
设计时间：2006—2007 年
建成时间：2009 年
项目功能：基础设施
业　　主：香港机场管理局
主要设计人：江立文，唐宙行

Location: Hong Kong, PRC
Gross floor area: 18,900 m^2
Design time: 2006-2007
Completion year: 2009
Section: Infrastructure
Client: Airport Authority Hong Kong
Directors: Max CONNOP, Albert TONG

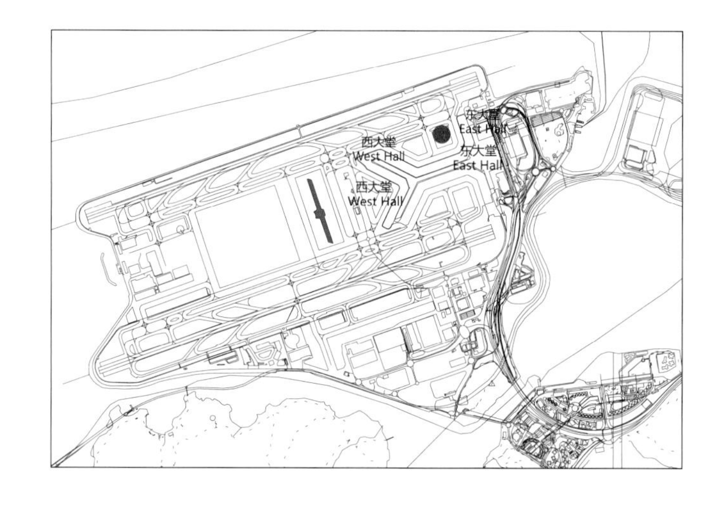

常规扁平化的屋顶结构，设计师特意设计了一个“航天舱”，将机电设施放置其中，让下方屋面形成波浪形状，既满足设备对空间的需求，又可将其巧妙地收藏于大厅上方。由于重量集中于屋顶中央区，因此客运廊内部将截面较大的承重柱对位排布在中部；翼侧相对轻盈，两侧区域的柱径明显纤细，富有装饰意味。线条的轻重对比更体现出内部空间灵动的变化。

空间：心理的感知

某种意义上而言，机场是极为特殊的空间，创造它不为留住人群，只为人们稍作停留而后离开——或入境市区，或去到更远的地方。Aedas 多年来设计了多座具有高辨识度的机场建筑，其特色之一是对旅客心理感受的敏锐觉察，并通过建筑设计的语言与手法将解决方案融入到空间中。在北卫星客运廊近窗的一侧，其曲面屋顶由“V”形钢管结构支撑，倾斜的结构有效受力，让旅客在焦灼候机望向窗外时得以获得更加广阔的视野；倾斜的细柱带来趣味性与韵律感，比之枯燥寡味的垂直立柱更能舒缓紧张的情绪。

细节：隐匿的巧思

为稳定支撑北卫星客运廊的玻璃幕墙，在水平向横梁与垂直向钢柱的交汇点处需要一组构件连接。设计师匠心独具地将分割后的工字钢柱以“背靠背”的形式连结，从而形成横截面为十字形的竖框，配上极具曲线美感的精巧抛光铸件节点与横梁连接。这种细节设计颇为含蓄，形色匆忙的旅客几乎不会留意到，但正如烹饪美馔的天厨会将细微的香草碎屑零星洒在菜肴之上，虽不显山露水却着实提升了餐食的整体品质。建筑师以这种独特的材料与形式的外立面支架捕捉了自然光线，在室内创造了一个令人愉悦的氛围。

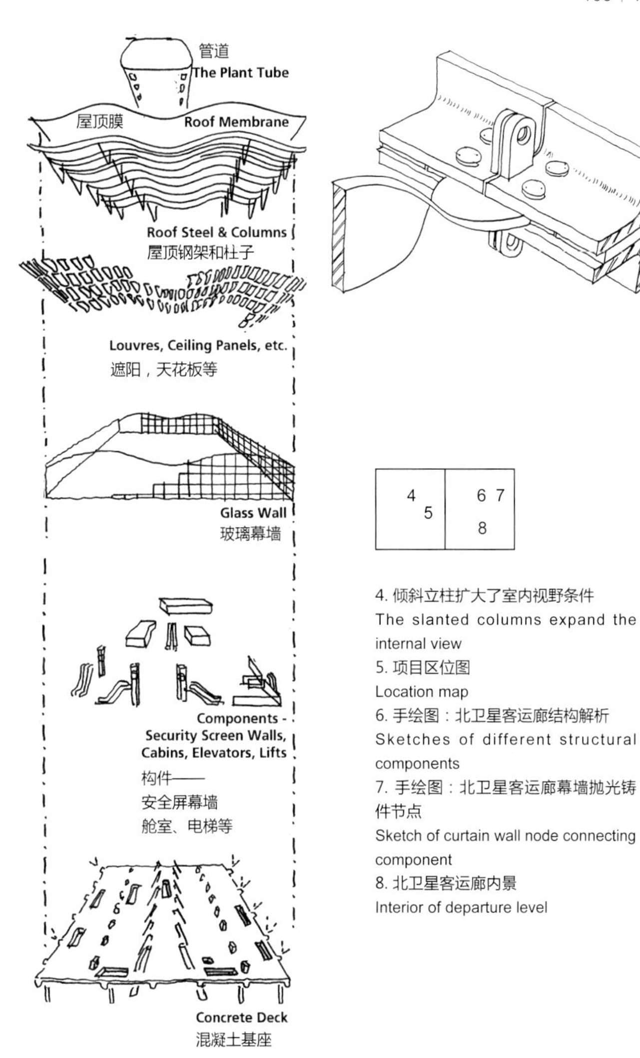

4 5 | 6 7 8

4. 倾斜立柱扩大了室内视野条件
The slanted columns expand the internal view
5. 项目区位图
Location map
6. 手绘图：北卫星客运廊结构解析
Sketches of different structural components
7. 手绘图：北卫星客运廊幕墙抛光铸件节点
Sketch of curtain wall node connecting component
8. 北卫星客运廊内景
Interior of departure level

The Hong Kong International Airport (HKIA) is located on Chek Lap Kok, next to Lantau Island in the New Territories of Hong Kong. Since it started serving the city in 1998, HKIA has been recognised as the world's best airport more than 60 times. In 2016, the number of passengers HKIA handled reached 70 million. HKIA North Satellite Concourse is a standalone fully operational passenger concourse with 10 bridge-served frontal stands and vehicle connection to existing Terminal 1. With a modern appearance and unique space, the North Satellite Concourse has become an integral part of what makes HKIA a great airport.

Form: Well-balanced

The dynamic winged roof features a rhythmic, flowing profile. The architectural form is derived from considerations of the complicated site conditions, building servicing, efficient structure and spatial needs. Due to the compact master layout, the land allocated for the concourse is very limited. If mechanical plant and equipment are placed under the roof, the roof would be much higher than allowed. Aedas ingeniously designed a "sky-pod" to house all the major plants and equipment, which was integrated into the wavy roof. Because the load is mainly in the central area of the roof, bigger columns are placed in the center of the concourse, and the perimeter props are accordingly smaller. The contrasting lines of the columns are aesthetically interesting, creating a distinctive interior space.

Space: Psychological Experience

Generally, airports are places to pass through temporarily, and people dislike staying there for long periods of time. Over the years, Aedas has designed several airport facilities. One of the features of those designs is the attention paid to travellers' psychological experience and how this relates to the airport's design. In the North Satellite Concourse, the curving roof structure is supported externally along the perimeter of the building by diagonal steel V-shaped tree columns, which enables visitors to enjoy a broader view of their surroundings, easing the stress of travel. Compared to conventional columns, the diagonal slender support is aesthetically more interesting.

Implicit Details

In order to effectively support the North Satellite Concourse's glass curtain wall, a set of structural elements are needed to connect horizontal beams and vertical steel columns. The support steel consists of a universal column spilt at the web and turned back to back to form a cruciform-section, and is connected to the transoms with a polished stainless steel twisting node connection. The detail is so implicit that may not be noticed by passengers who are in a hurry. Like the arrangement of a plate of food, it is subtle yet it makes a great impression. This design definitely brings positive impacts to the overall quality of the construction. The design used this unique material and form of the facade brackets to catch the light, creating a delightful effect in the interior.

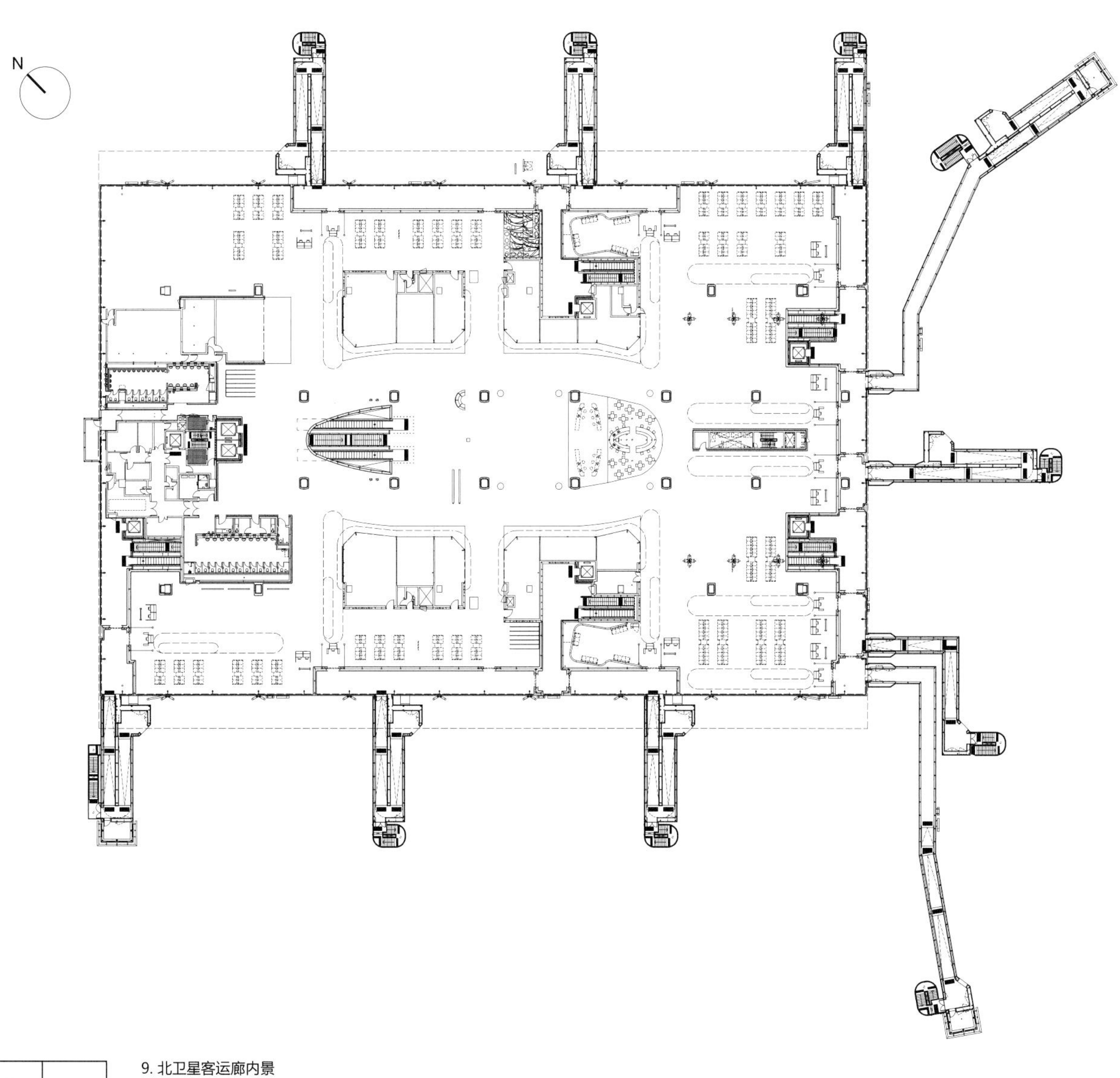

9	10

9. 北卫星客运廊内景
Interior of departure level
10. 一层平面图
Level 1 floor plan

香港富临阁

The Forum, Hong Kong

富临阁位于香港中环，这里被称为香港华尔街的中心，周边高楼林立，交通四通八达。业主香港置地希望对富临阁大厦实施改造，包括改善整个基地的连通性，并设置室外广场、新建一栋小型建筑，并摆放亨利·摩尔（Henry Moore）、朱铭（Ju Ming）和伊莉莎伯·弗林克（Elisabeth Frink）设计的室外雕塑。

新建的小型建筑，坐落于公交枢纽站及停车库上方。在鳞次栉比的高楼中，仅仅五层的富临阁尺度亲和，像在地上镶嵌了一颗静谧的宝石，又如遗世的沧海明珠。建筑落成时渣打银行正致力于品牌化建设，希望以一些特别方式来吸引社会的关注，因此当原租户搬出，渣打银行便着手将总部迁入。

重新定义的富临阁被视作一个“倾斜的盒子”，方形的体块和倾倒15°的构思为建筑赋予了功能之外的意义，低矮的建筑成为地标，缓冲了密集的超高层建筑群形成的空间压迫。倾斜的外立面是钢质斜肋结构，玻璃幕墙透出交织的白色骨骼，幕墙反射出周边建筑并将倒影切割成均匀的片段。倾斜的建筑体量在顶部伸出一角形成花园，在底部支起一角形成入口，办公写字楼收纳在中央。屋顶花园由上至下，层层叠落，为蓝宝石上新添了许多绿意。

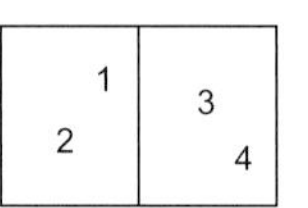

1. 香港中环高楼林立，被称为“香港华尔街的中心”
Central, the Hong Kong version of Wall Street
2. 富临阁入口
Entrance of The Forum
3. 富临阁在视觉上仿佛一个“倾斜的盒子”
The Forum: an oblique box
4. 项目区位图
Location map

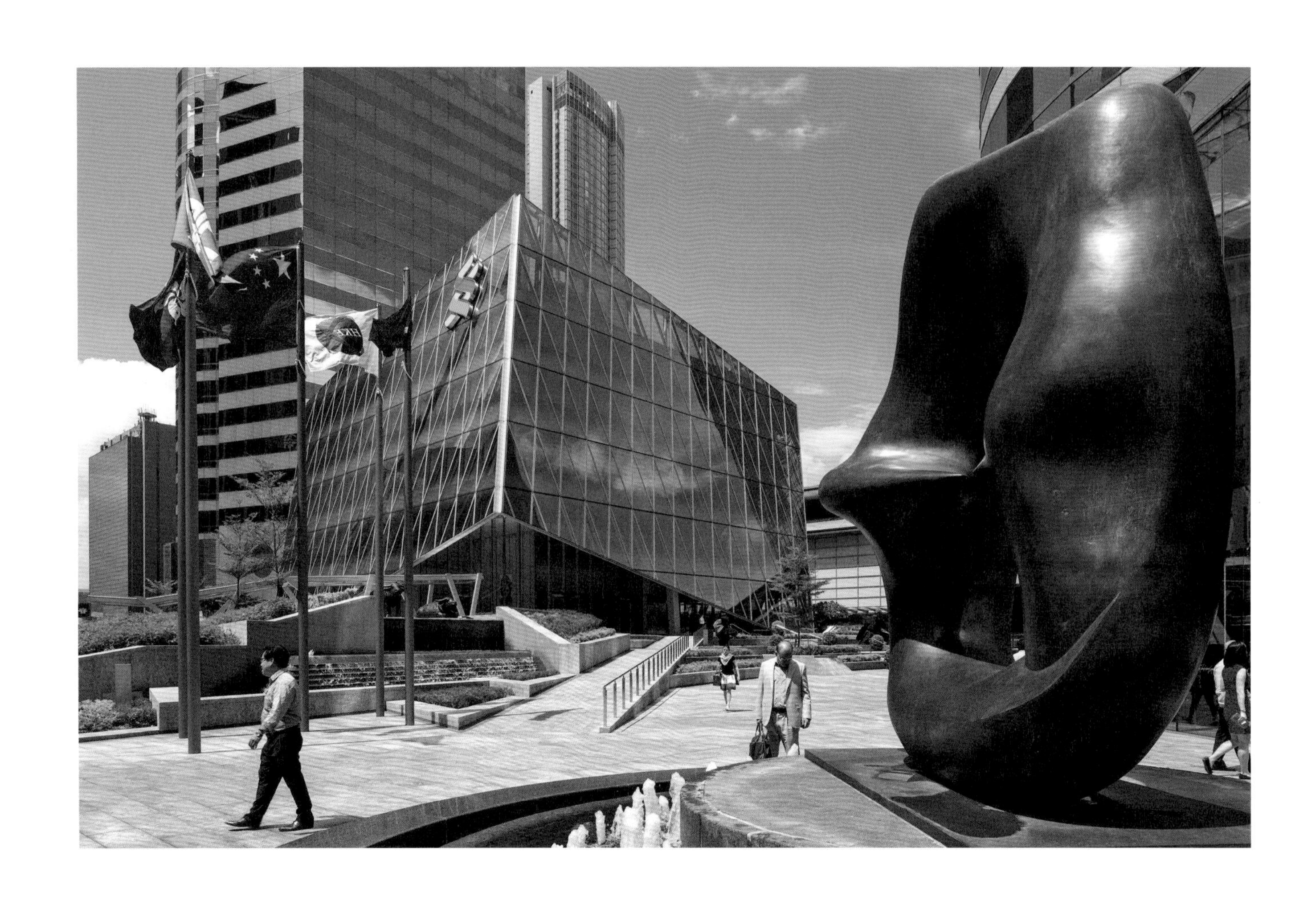

项目信息 | PROJECT INFORMATION

地　　点：中国香港
建筑面积：4 501m²
建成时间：2014 年
项目功能：办公楼
业　　主：置地控股有限公司
主要设计人：纪达夫

Location: Hong Kong, PRC
Gross floor area: 4,501 m^2
Completion year: 2014
Sector: Office
Client: Hongkong Land Limited
Director: Keith GRIFFITHS

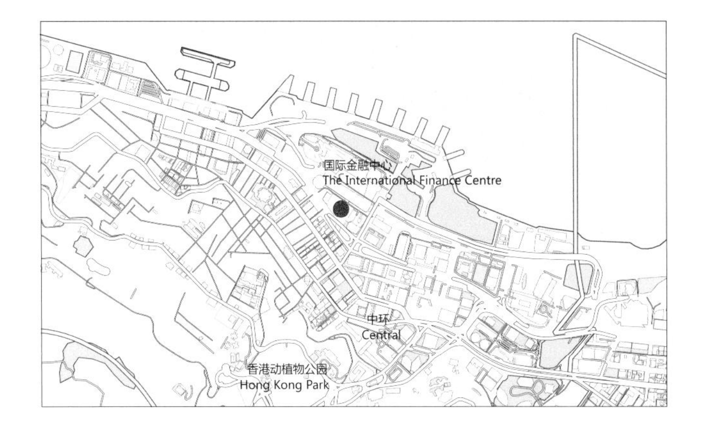

The Forum is located in the heart of Central, Hong Kong's main business district and the Asian answer to Wall Street. The property owner and client, Hong Kong Land, wanted to transform the site by offering better connectivity with the surrounding area. That involved developing an outdoor plaza, putting up a small office building and deciding on new locations for some outstanding outdoor sculptures by Henry Moore, Ju Ming and Elisabeth Frink. The new small-scale architecture, located above a public transportation hub, contains five storeys, and it resembles as a gem or pearl mounted in a crown. Standard Chartered Bank, was in the middle of a branding exercise and needed something special as a corporate focus. When the original tenant pulled out, the bank stepped in and decided to move its headquarters staff into The Forum.

The new Forum is a tilted box. The cubic mass and the 15 degree tilt add visual interest as well as functionality. The small building has become a landmark and it also relieved the congested urban space that surrounds it. The building's glass curtain wall is framed by steel mullions and supported by white-coloured structures, reflecting the surrounding buildings in a mosaic pattern. The tilted box has a garden on the rooftop and contains an entrance on the lower level that provides access to the offices. The terraced rooftop garden adds an amazing green touch to this blue jewel.

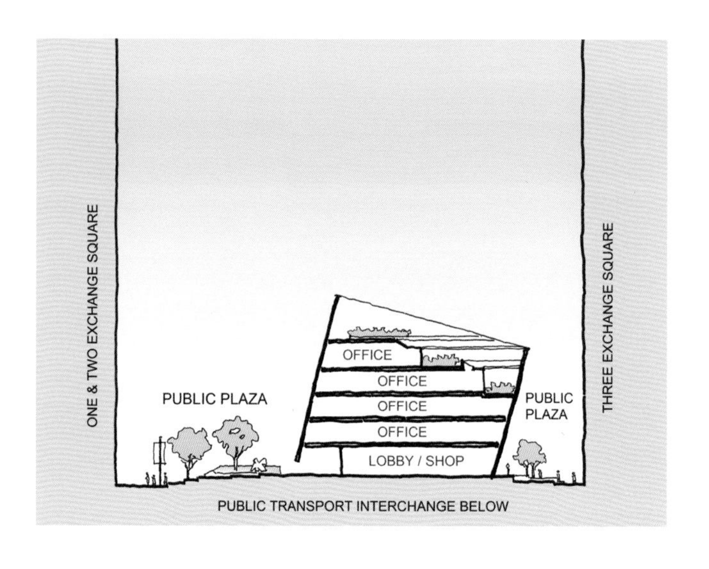

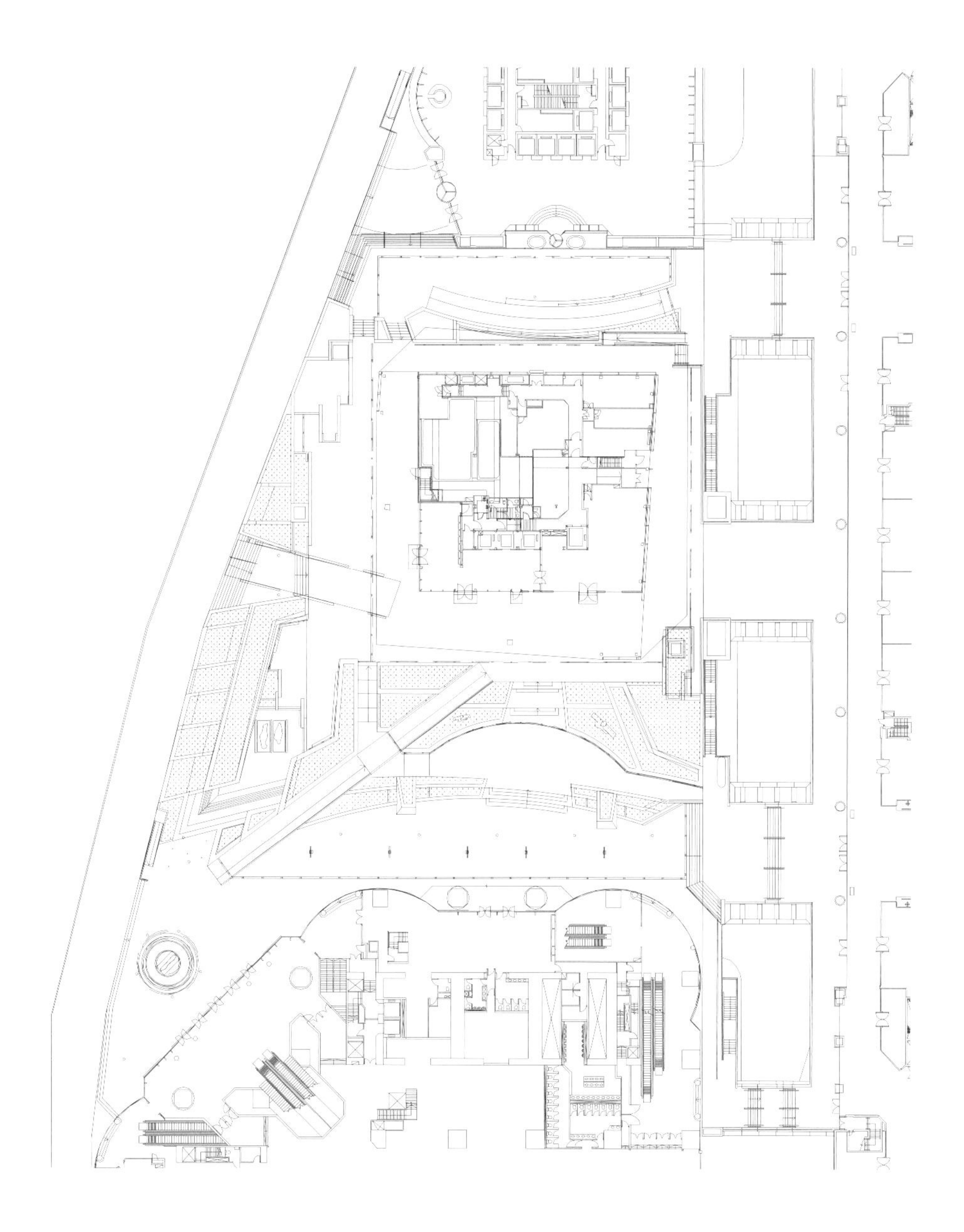

5	7
6	8

5. 俯瞰富临阁
Aerial view
6. 剖面图
Section
7. 首层景观总平面
Level 1 floor plan
8. 立面图
Elevation

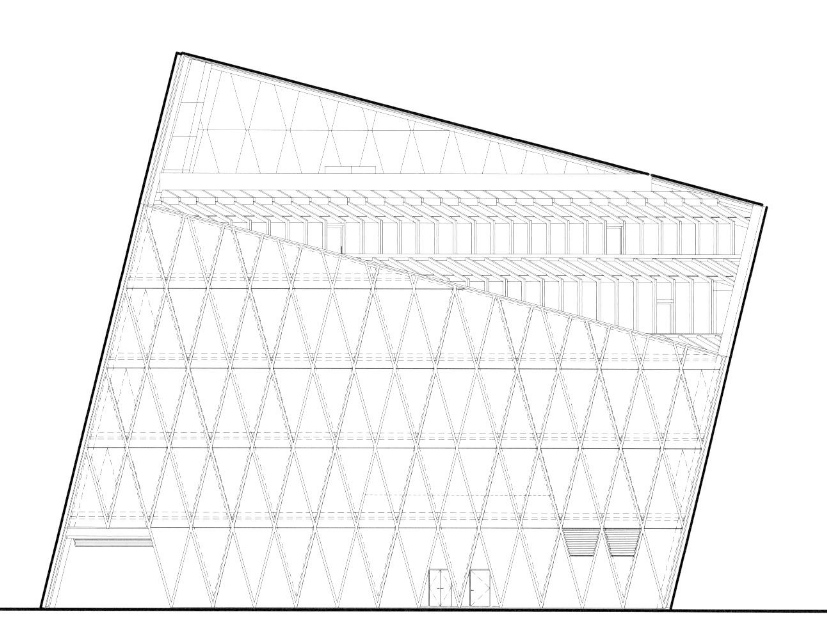

台北砳建筑

Lè Architecture, Taipei

位于中国台湾地区基隆河南岸的砳建筑恰如其名“砳”，仿佛河畔一高一低两块混沌未明的“卵石”，在经历时间涤荡后变得光洁圆润，展现出封闭、内敛、含蓄的气质。

台北当地的气候对遮阳的要求很高，尤其是西晒，建筑设计在西面广阔的百叶幕墙上设置了许多倾斜的网架，安置了不同种类的植物在建筑表皮中生长呼吸。在建筑公共空间的功能规划上，南面设置了跨层的露台，它是面对城市布置的“城市客厅”，为在这里工作的人们提供了一个公共交流与活动的场所，几个楼层的人都可以到这里停留、休憩，在露台上饱览台北之景。作为“城市客厅”，露台是建筑对外舒展自我的部分，它使建筑与城市达成融合，而非孤立的个体。从结构角度来看，砳建筑最夺目的莫过于入口处轻盈庞大的出挑雨棚，轻薄的雨棚将两块“卵石”连为一体。雨棚采用了悬挑的结构，结构“骨骼”像横置的鱼骨，覆盖在底层空间的上方。

1. 台北城市夜景
Night view of Taipei
2. 位于基隆河南岸的砳建筑
Lè Architecture on the southern bank of Keelung River
3. 砳建筑东南向立面
Southeast elevation

项目信息 | PROJECT INFORMATION

地　　点：中国台北
建筑面积：11 449.8m^2
设计时间：2011—2015 年
建成时间：2017 年
项目功能：办公楼
业　　主：诚意开发股份有限公司
主要设计人：温子先博士

Location: Taipei, PRC
Gross floor area: 11,449.8 m^2
Design time: 2011-2015
Completion year: 2017
Sector: Office
Client: Earnest Development & Construction Corporation
Director: Dr. Andy WEN

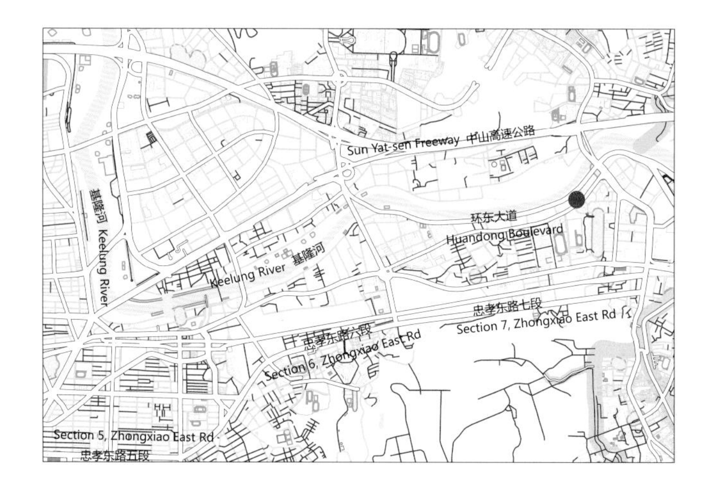

Located in close proximity to the Keelung River in Taipei, this project consists of two buildings inspired by river pebbles. Just like the stones polished by time, the buildings convey a moderate, introverted personality.

Taipei's climate calls for buildings to have shading devices installed, especially on the west-facing elevations, and here is a green planter system that serves the purpose by acting as a facade filter. There are cross floor decks, a kind of "urban living room", to provide the office workers a public space to chat, rest and enjoy the view of Taipei. The decks act as an interface where the building meets the city and becomes an integral part of the surrounding environment. In terms of structure, the most eye-catching part in Lè Architecture is the enormous overhanging canopy above the entrance. The light, thin fish-bone like structure brings the two "stones" together.

4 5	6 7 8

4. 砳建筑全景
External view
5. 项目区位图
Location map
6.7. 手绘图
Sketches
8. 砳建筑轻盈出挑的入口雨棚，将两部分建筑连为一体
The canopy that connects the two buildings

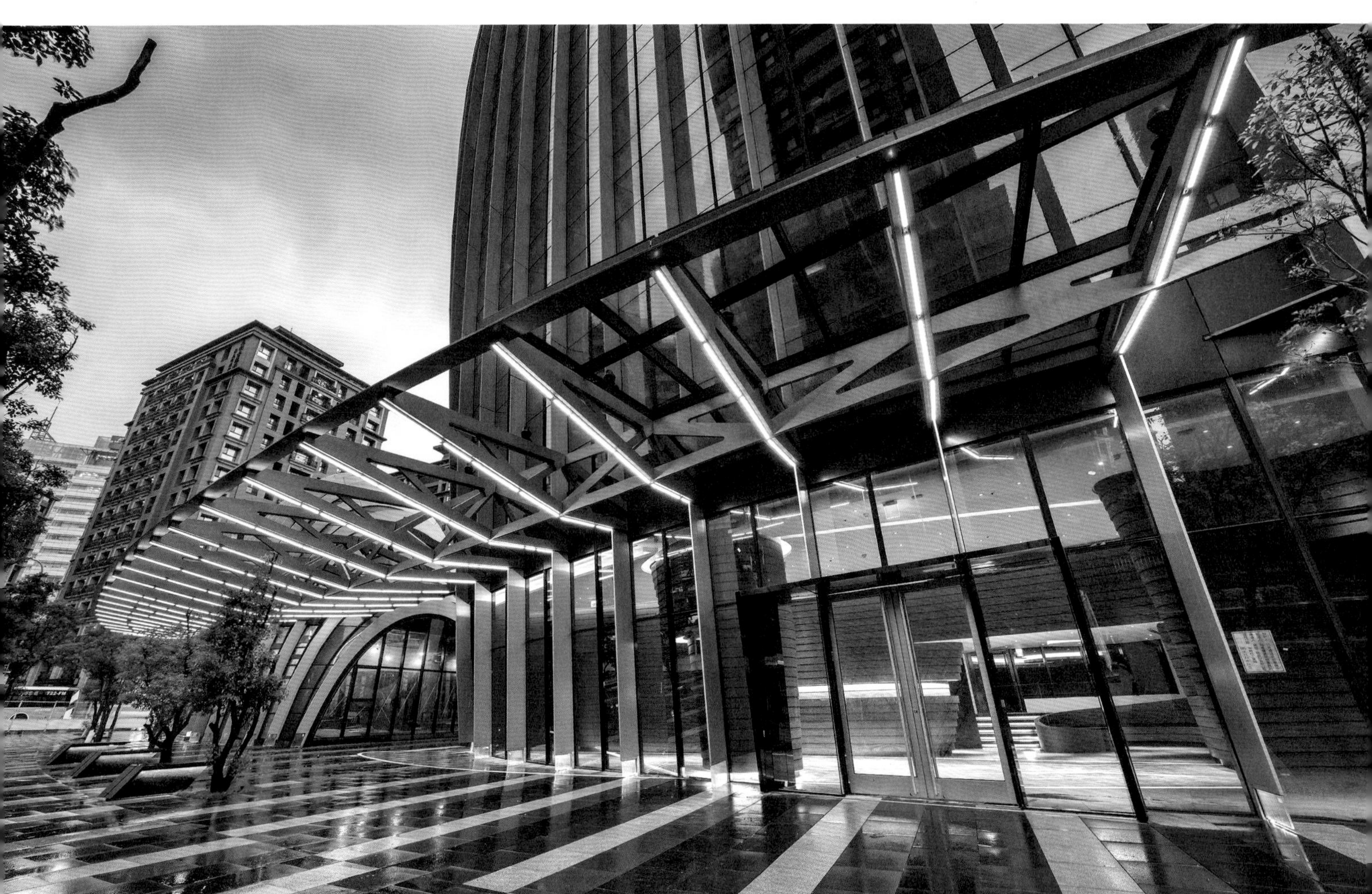

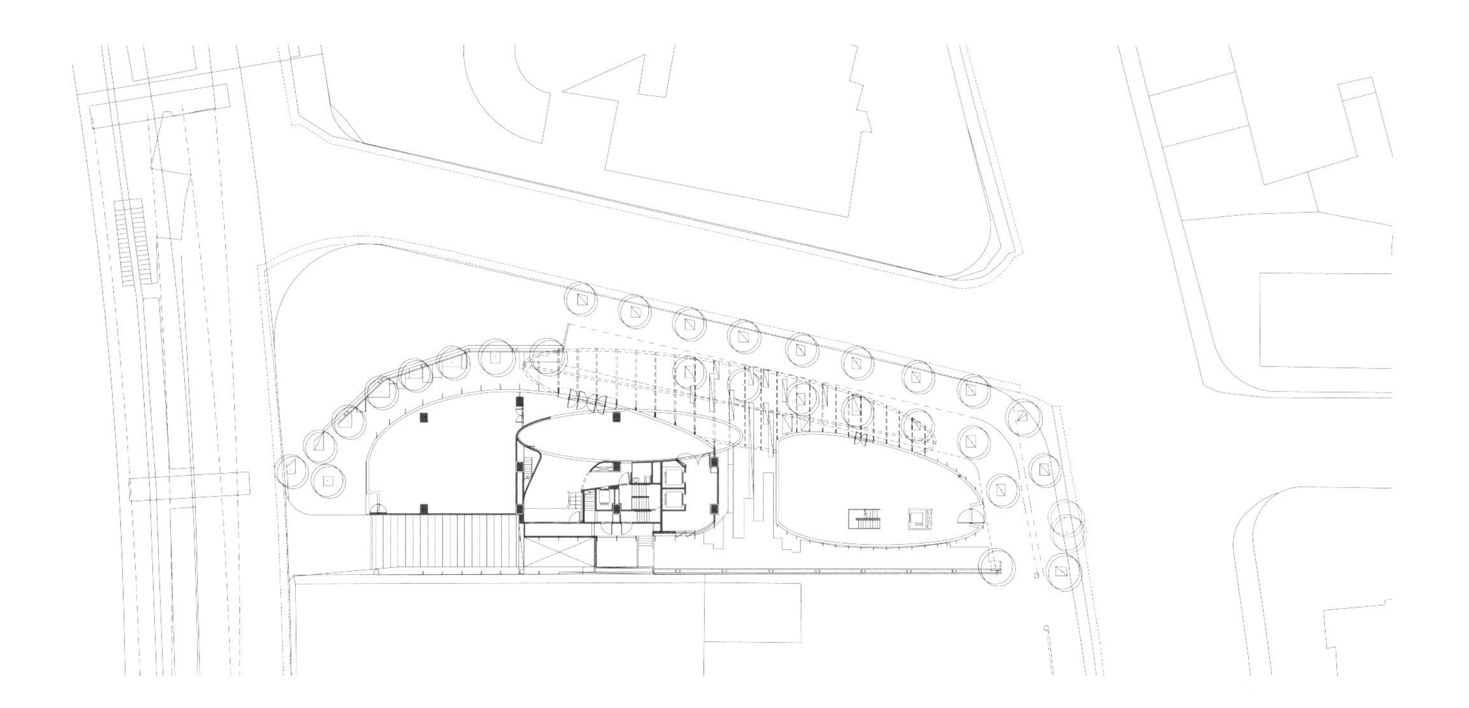

9	11
10	12

9. 砳建筑整体鸟瞰
Aerial view
10. 砳建筑内部
Lobby area
11. 一层平面图
Level 1 floor plan
12. 西南立面图
Southwest elevation

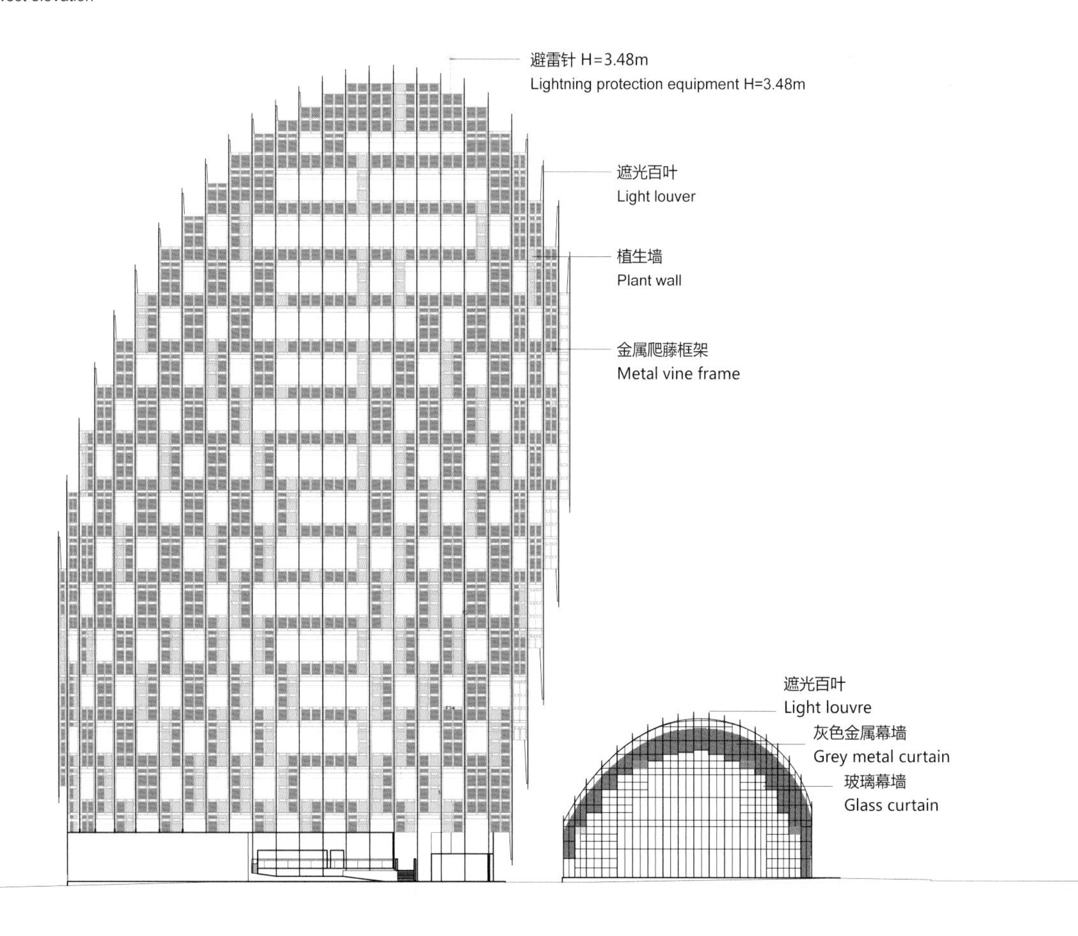

台中商业银行企业总部综合项目

Commercial Bank Headquarters Mixed-use Project, Taichung

台中是中国台湾地区的第二大城市，也是台湾中部的经济、交通、文化中心。台中商业银行企业总部综合项目位于市中心西屯区的市政路与河南路交汇处，毗邻歌剧院和夏绿地公园。

项目的设计概念以台中商业银行标志为基础，源自汉字“中”。设计避免将所有大型功能区堆叠在单一塔楼中的做法，而是打造了两座“分离”的塔楼，并在其间设置垂直空间以容纳大型设施。塔楼高 200m，包含 23 000m^2 的台中商业银行企业总部及 43 600m^2 的国际五星级酒店。一系列轻盈的透明玻璃盒子“漂浮”于两座塔楼间的空间中，容纳了例如公共展览空间、空中花园、宴会厅、游泳池及会议厅等大型功能设施。这种做法不仅使建筑外形更加丰富，还为该区域打造了一座面向城市主干道的独特的标志性建筑。屋顶的露台设有餐厅和 VIP 俱乐部，连同室外绿植露台，为顾客提供了引人注目的阳台空间和壮丽的城市景观。

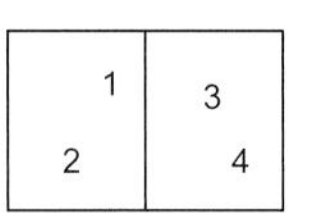

1. 繁华的台中东海夜市
Bustling Tunghai Night Market in Taichung
2. 丰富的建筑形体
A rich architectural form
3. 台中商业银行企业总部是一座面向主干道的独特标志性建筑
The building is a unique and iconic architecture facing the main road
4. 项目区位图
Location map

项目信息 | PROJECT INFORMATION

地 点：中国台中
建筑面积：110 000m^2
设计时间：2018—2020 年
建成时间：预计 2020 年
项目功能：综合体
业　　主：台中银行及余晓岚建筑师事务所
主要设计人：纪达夫

Location: Taichung, PRC
Gross floor area: 110,000 m^2
Design time: 2018-2020
Completion year: 2020
Sector: Mixed-use
Client: Taichung Bank and Y.S.L. Architect & Associates
Director: Keith GRIFFITHS

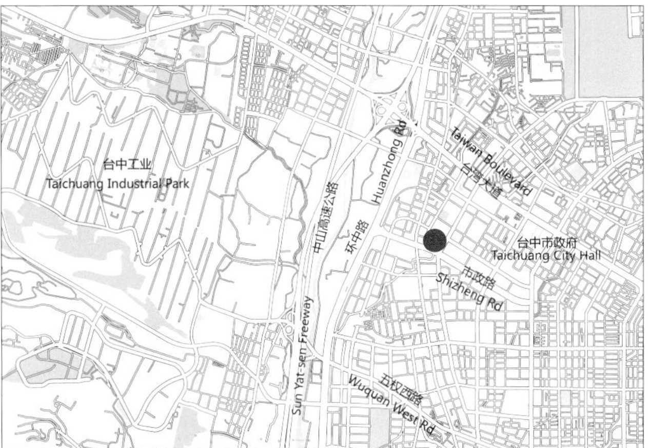

禮客
OUTLET
LEECO
LEECO
購

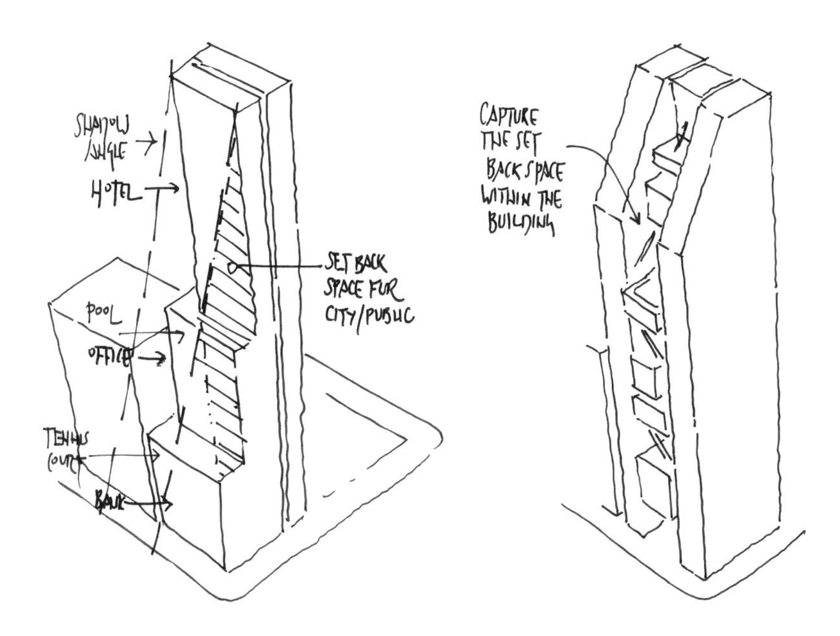

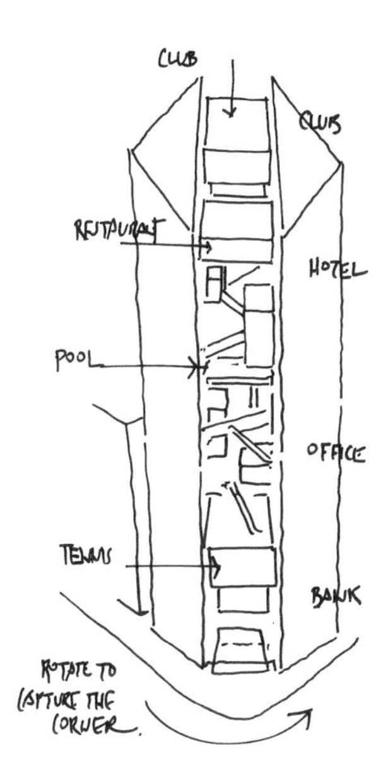

The design concept for this project originates from the logo of Taichung Commercial Bank, which contains the Chinese character "中", meaning "center." Instead of stacking all the major functions in one singular tower, the design creates two "separate" towers with a vertical void in the middle to accommodate a number of large facilities.

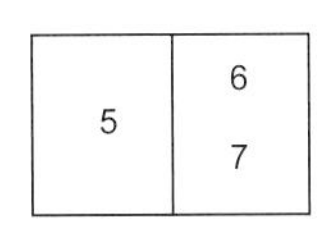

5. 远观台中商业银行企业总部综合项目
Overview of the project
6. 手绘图
Sketches
7. 轻盈的玻璃盒子容纳了一系列大型功能设施
Light glass boxes house a series of large-scale functional facilities

The 200-metre towers include 23,000 square metres space for the Taichung Commercial Bank Headquarters and 43,600 square metres of space for an international five-star hotel. Between the two towers, a series of light and transparent glass boxes appear to "float" inside the void to house large-scale functional facilities including public exhibition space, sky gardens, a ballroom, swimming pool and conference facilities.

This design has also created a unique and iconic building facing Taichung's main road. The landscaped rooftop terrace features a restaurant, VIP club as well as a spectacular observation deck for visitors.

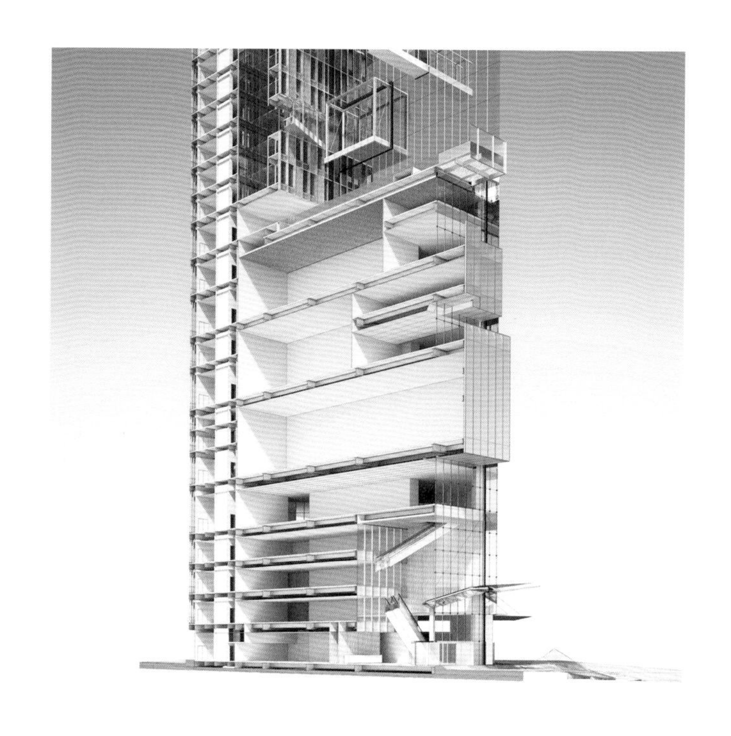

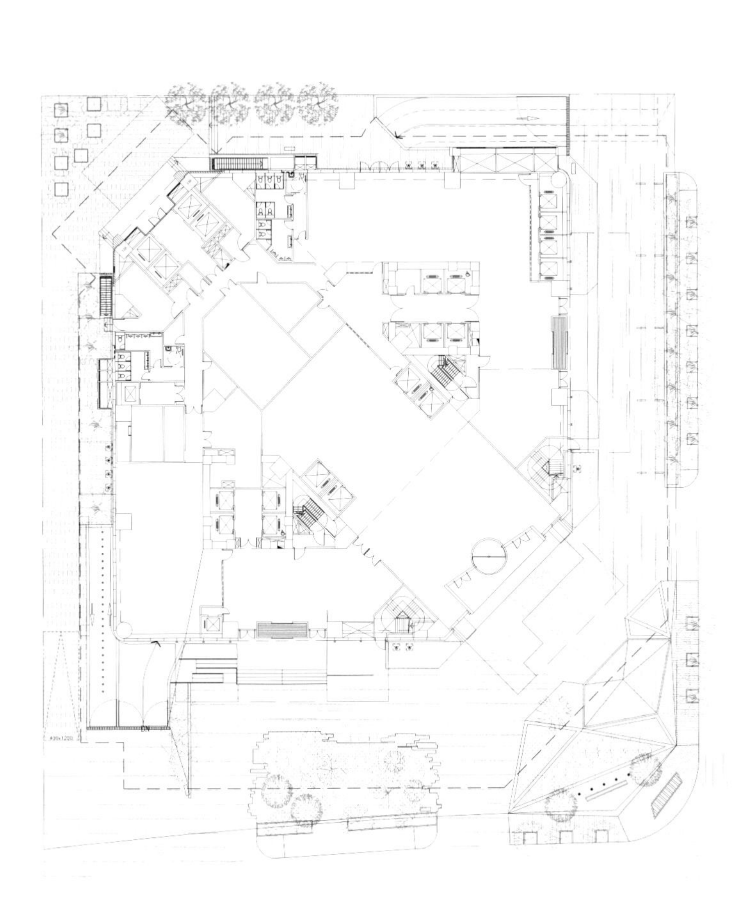

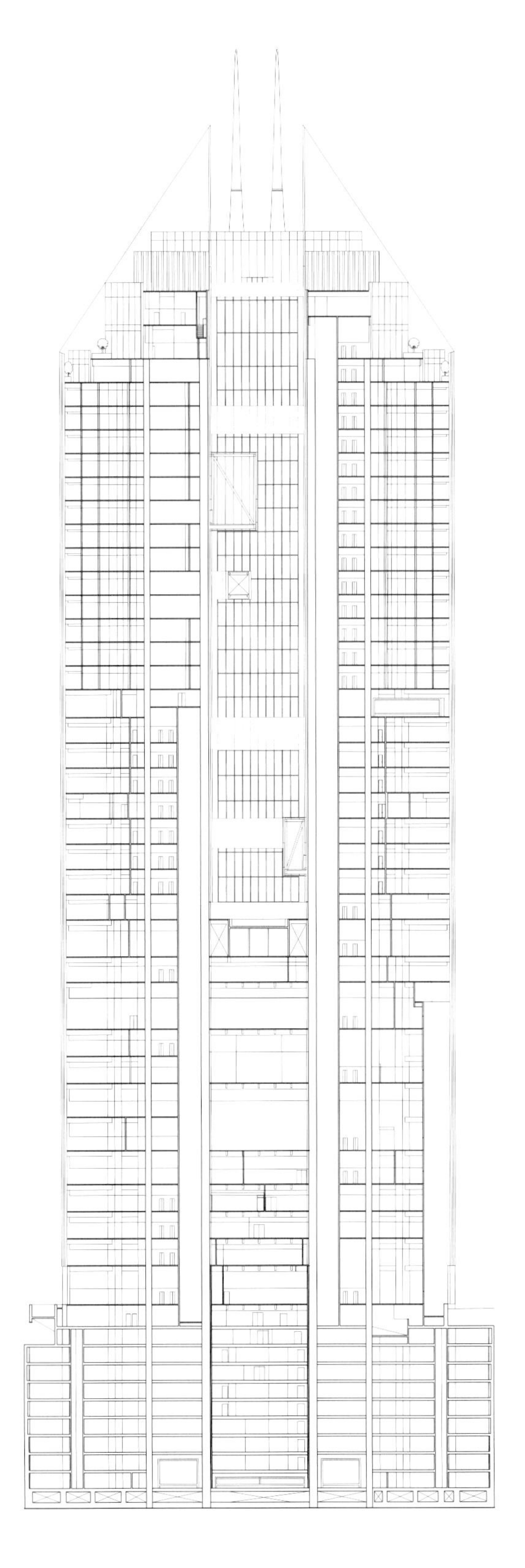

8 9	10 11

8. 分析图
Analysis diagram
9. 空中泳池
Sky pool
10. 一层平面图
Level 1 floor plan
11. 剖面图
Section

一带一路
BELT AND ROAD

中国大陆长江以北 NORTH OF YANGTZE RIVER

- ❷❸❹❼ 大连恒隆广场
Olympia 66, Dalian
- ❷❹❻ 北京大望京综合开发项目
Da Wang Jing Mixed-use Development, Beijing
- ❸❹ 北京大兴 3 及 4 地块项目
Daxing Plots 3 and 4, Beijing
- ❶❸❻ 北京新浪总部大楼
Sina Plaza, Beijing
- ❶❷❸❹❻ 北京北苑北辰综合体
North Star Mixed-use Development, Beijing
- ❷❸❹❼ 青岛金茂湾购物中心
Jinmao Harbour Shopping Center, Qingdao
- ❶❷❹ 成都恒大广场
Evergrande Plaza, Chengdu
- ❶❼ 重庆新华书店集团公司解放碑时尚文化城
Xinhua Bookstore Group Jiefangbei Book City Mixed-use Project, Chongqing
- ❷❸❹❼ 武汉恒隆广场
Heartland 66, Wuhan

中国大陆长江以南 SOUTH OF YANGTZE RIVER

- ❸❻❼ 上海星荟中心
Shanghai Landmark Center, Shanghai
- ❶❷❸❹❻ 上海龙湖虹桥项目
Longfor Hongqiao Mixed-use Project, Shanghai
- ❻❼ 上海虹桥世界中心
Hongqiao World Center, Shanghai
- ❷❸❹❻❼ 无锡恒隆广场
Center 66, Wuxi
- ❸❼ 苏州西交利物浦大学中心楼
Xi'an Jiaotong-Liverpool University Central Building, Suzhou
- ❷❸ 义乌之心
The Heart of Yiwu, Yiwu
- ❷❸❻ 广州南丰商业、酒店及展览综合大楼
Nanfung Commercial, Hospitality and Exhibition Complex, Guangzhou
- ❸❽ 广州邦华环球贸易中心
Bravo PARK PLACE, Guangzhou
- ❸ 珠海粤澳合作中医药科技产业园总部大楼
Headquarters, Traditional Chinese Medicine Science and Technology Industrial Park of Co-operation between Guangdong and Macao, Zhuhai
- ❸❽ 深圳宝安国际机场卫星厅
Shenzhen Airport Satellite Concourse, Shenzhen
- ❶❼ 珠海横琴国际金融中心
Hengqin International Financial Center, Zhuhai
- ❷❸❹ 珠海横琴中冶总部大厦（二期）
Hengqin MCC Headquarters Complex (Phase II), Zhuhai

“一带一路”是中国提出的全球融合发展的合作倡议，以共商、共建、共享为宗旨，将中国改革开放四十年的经验模式与沿线各国分享，并发展经济合作伙伴关系，同时促成中国与一带一路沿线在基础设施、城市建设、商业贸易、文化交流等多方面的合作。Aedas 为这些国家、地区和城市的建设留下了一些特色的建筑实践，这样的交流融合将给这些地区带来新的机遇与发展，分享中国经济发展的价值，共享全球和谐发展。

The Belt and Road is a new initiative introduced by China to share its 40-year development experience with countries along the land and sea routes. It enables collaborations in infrastructure, urban development, trading and cultural exchanges. Aedas has been bringing tailored architecture to these cities. Such exchanges and integration will bring new opportunities and development to these regions, share the value of China's economic development, and facilitate global development.

中国香港地区和中国台湾地区 HONG KONG AND TAIWAN REGIONS OF CHINA

1 5 8 香港西九龙站
Hong Kong West Kowloon Station, Hong Kong

2 3 5 6 港珠澳大桥香港口岸旅检大楼
Hong Kong-Zhuhai-Macao Bridge Hong Kong Port - Passenger Clearance Building, Hong Kong

1 5 8 香港国际机场中场客运廊
Hong Kong International Airport Midfield Concourse, Hong Kong

1 5 8 香港国际机场北卫星客运廊
Hong Kong International Airport North Satellite Concourse, Hong Kong

4 香港富临阁
The Forum, Hong Kong

3 7 台北砳建筑
Lè Architecture, Taipei

3 台中商业银行企业总部综合项目
Commercial Bank Headquarters Mixed-use Project, Taichung

一带一路 BELT AND ROAD

3 4 8 新加坡星宇项目
The Star, Singapore

3 8 新加坡 Sandcrawler
Sandcrawler, Singapore

8 阿联酋迪拜 Ocean Heights
Ocean Heights, Dubai, UAE

5 8 阿联酋迪拜地铁站
Dubai Metro, Dubai, UAE

8 9 英国唐卡斯特 Cast 剧院
Cast, Doncaster, UK

九项设计理念 NINE POINTS OF DESIGN IDEAS

1 《国家新型城镇化规划（2014—2020 年）》
NATIONAL NEW URBANISATION PLAN (2014-2020)

2 高密度城市枢纽
HIGH DENSITY CITY HUBS

3 具有通透性和连接性的设计
POROUS AND CONNECTED DESIGNS

4 新型商业零售
THE NEW RETAIL

5 基础设施设计
INFRASTRUCTURE DESIGN

6 中国的特大城市与城市设计
CHINA MEGAPOLIS AND URBAN DESIGN

7 文化和故事
CULTURE AND STORY

8 一带一路
BELT AND ROAD

9 城市改造与修复
URBAN RENEWAL AND RESTORATION

新加坡星宇项目

The Star, Singapore

活力狮城的波那维斯达区，是新加坡最重要的科研中心和教育重镇，区内有著名的纬壹科技城、新加坡国立大学、新加坡理工学院等科教机构。星宇项目位于横穿波那维斯达区的一处狭长场地，这里的地文气候特征十分明显。为顺应区位环境，Aedas 以柔和、流动的设计手法贯穿整个项目，让建筑项目与花园之城相融合。

星宇项目主要包括一个可容纳5 000个座位的剧院和含有餐饮的多元化商业体。大剧院位于 40m 的高空，由多根倾斜的混凝土支柱支撑，剧场下方是由商铺及餐厅组成的商业楼层。为应对复杂的高差问题，设计师在不同区域分别设置了两个入口层，不仅适应标高上的变化，也将落客点融入到开放式广场中，吸引游客从各个方向进入星宇。35m 通高的大型广场立于中央，大大拓展了水平和垂直两个方向的空间及视觉流通。此外，被动式节能是又一设计亮点，为改善闷热的气候，建筑特别设计了被动式通风系统，利用入口的高差实现空气的流动。Aedas 对日光、太阳辐射、眩光等因素进行精确分析，为调节项目内外的微气候做出了行之有效的环境提升。

1 2 3

1. 星宇项目充满张力的造型正应和了新加坡的城市活力
The form of The Star demonstrates the vibrancy of the Lion City
2. 开放广场为各个方向前来的行人提供接近星宇项目的渠道
The open plaza allows access from different directions
3. 星宇项目外景
External view

CAR PARK ENTRANCE

4 5	6 7

4. 多层次的设计与山地的基地背景相结合
The multi-level design adapts to the site's hilly terrain
5. 项目区位图
Location map
6. 功能空间分析示意图
Function analysis diagram
7. 倾斜且灵活安置的立柱为商场空间增添了活力
Playfully slanted columns add vitality to the mall

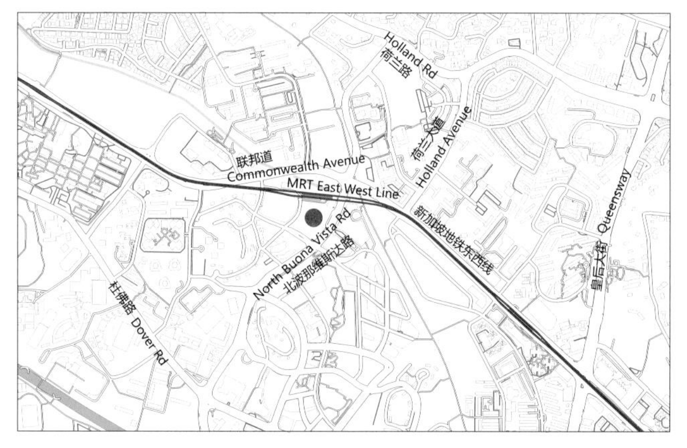

项目信息 | PROJECT INFORMATION

地　　点：新加坡纬壹
建筑面积：62 000m²
设计时间：2007—2012 年
建成时间：2012 年
项目功能：综合体
业　　主：Rock Productions Pte Ltd 及凯德商用产业有限公司
主要设计人：Andrew BROMBERG

Location: One-north, Singapore
Gross floor area: 62,000 m²
Design time: 2007-2012
Completion year: 2012
Sector: Mixed-use
Client: Rock Productions Pte Ltd., CapitaMalls Asia Ltd.
Design Director: Andrew BROMBERG

Bouna Vista is the most important research and education district in Singapore — the Lion City — and it is home to the famous one-north district, the National University of Singapore and the Singapore Polytechnic. The Star is located in a long and narrow site in Bouna Vista that has distinctive geo-climatic features. In response to the environment, Aedas designed a project with a soft and flowing expression that fits into the atmosphere of the garden city.

The Star consists of a 5,000-seat theatre and a diversified retail complex that includes food and beverage outlets. The vast auditorium is perched 40 meters up in the air on a forest of slanting concrete pillars. Below the theatre are several floors of shops and

CAR PARK ENTRANCE

4	6
5	7

4. 多层次的设计与山地的基地背景相结合
The multi-level design adapts to the site's hilly terrain
5. 项目区位图
Location map
6. 功能空间分析示意图
Function analysis diagram
7. 倾斜且灵活安置的立柱为商场空间增添了活力
Playfully slanted columns add vitality to the mall

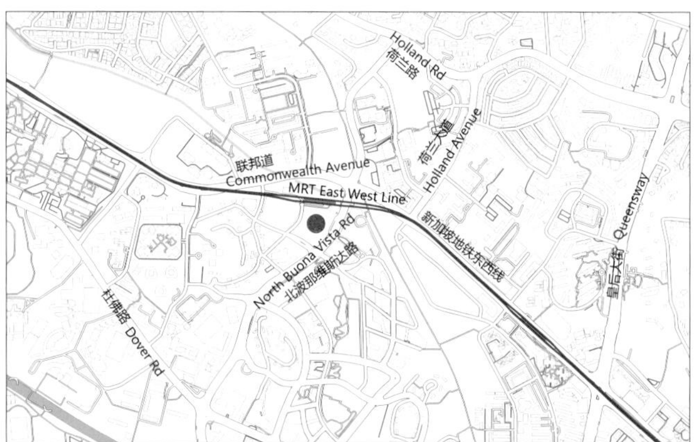

项目信息 | PROJECT INFORMATION

地　　点：新加坡纬壹
建筑面积：62 000m²
设计时间：2007—2012 年
建成时间：2012 年
项目功能：综合体
业　　主：Rock Productions Pte Ltd 及凯德商用产业有限公司
主要设计人：Andrew BROMBERG

Location: One-north, Singapore
Gross floor area: 62,000 m²
Design time: 2007-2012
Completion year: 2012
Sector: Mixed-use
Client: Rock Productions Pte Ltd., CapitaMalls Asia Ltd.
Design Director: Andrew BROMBERG

Bouna Vista is the most important research and education district in Singapore — the Lion City — and it is home to the famous one-north district, the National University of Singapore and the Singapore Polytechnic. The Star is located in a long and narrow site in Bouna Vista that has distinctive geo-climatic features. In response to the environment, Aedas designed a project with a soft and flowing expression that fits into the atmosphere of the garden city.

The Star consists of a 5,000-seat theatre and a diversified retail complex that includes food and beverage outlets. The vast auditorium is perched 40 meters up in the air on a forest of slanting concrete pillars. Below the theatre are several floors of shops and

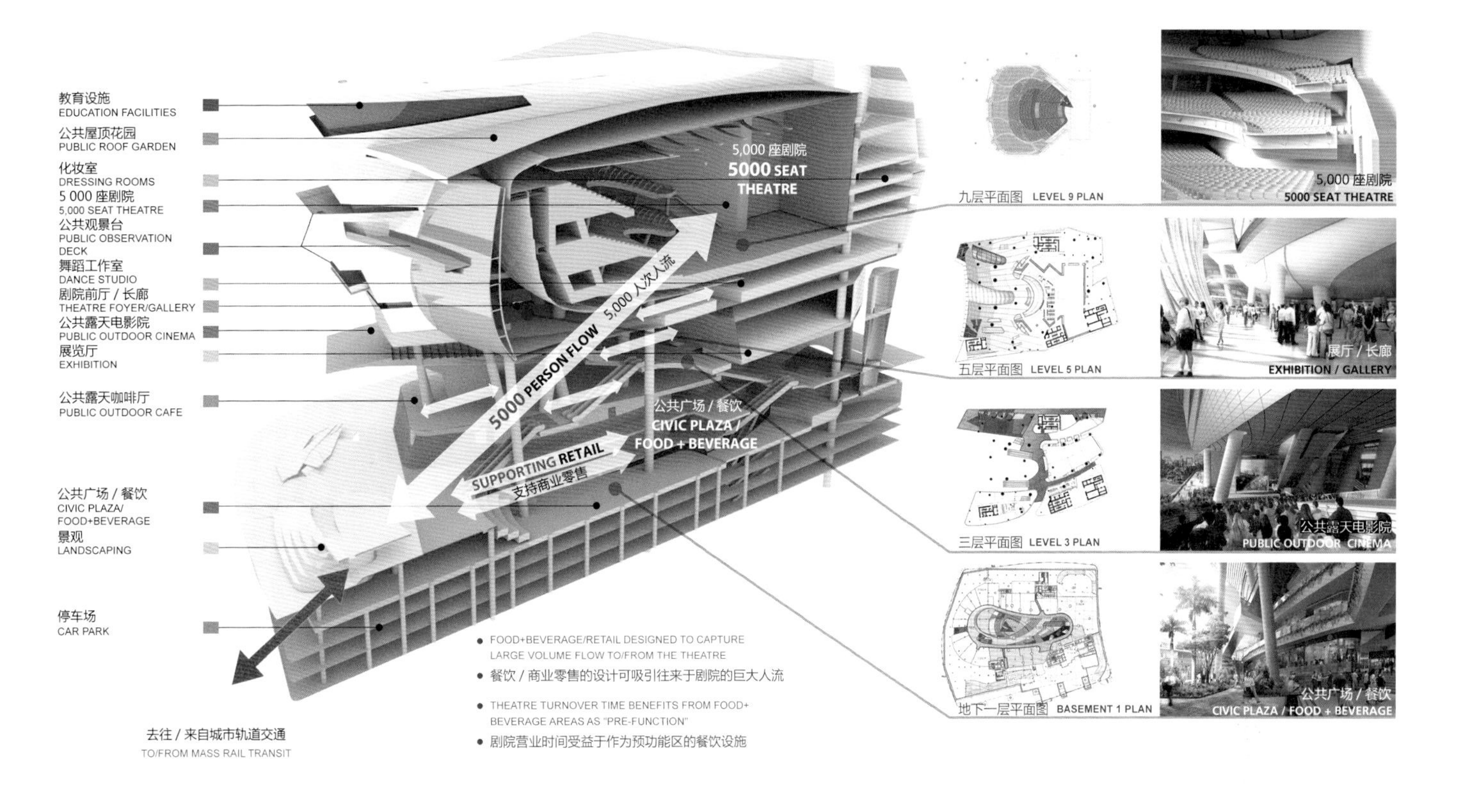
教育设施
EDUCATION FACILITIES
公共屋顶花园
PUBLIC ROOF GARDEN
化妆室
DRESSING ROOMS
5 000 座剧院
5,000 SEAT THEATRE
公共观景台
PUBLIC OBSERVATION DECK
舞蹈工作室
DANCE STUDIO
剧院前厅 / 长廊
THEATRE FOYER/GALLERY
公共露天电影院
PUBLIC OUTDOOR CINEMA
展览厅
EXHIBITION
公共露天咖啡厅
PUBLIC OUTDOOR CAFE
公共广场 / 餐饮
CIVIC PLAZA/ FOOD+BEVERAGE
景观
LANDSCAPING
停车场
CAR PARK
5,000 座剧院
5000 SEAT THEATRE
5000 PERSON FLOW
5,000 人次人流
公共广场 / 餐饮
CIVIC PLAZA / FOOD + BEVERAGE
SUPPORTING RETAIL
支持商业零售
去往 / 来自城市轨道交通
TO/FROM MASS RAIL TRANSIT
FOOD+BEVERAGE/RETAIL DESIGNED TO CAPTURE LARGE VOLUME FLOW TO/FROM THE THEATRE
餐饮 / 商业零售的设计可吸引往来于剧院的巨大人流
THEATRE TURNOVER TIME BENEFITS FROM FOOD+ BEVERAGE AREAS AS "PRE-FUNCTION"
剧院营业时间受益于作为预功能区的餐饮设施
九层平面图 LEVEL 9 PLAN
5,000 座剧院
5000 SEAT THEATRE
五层平面图 LEVEL 5 PLAN
展厅 / 长廊
EXHIBITION / GALLERY
三层平面图 LEVEL 3 PLAN
公共露天电影院
PUBLIC OUTDOOR CINEMA
地下一层平面图 BASEMENT 1 PLAN
公共广场 / 餐饮
CIVIC PLAZA / FOOD + BEVERAGE

dining outlets. A series of staircases and escalators connects the retail and cultural elements. In response to the complicated issue of height differences, the architect designed two entrance levels and located drop-off points in the open plaza to attract visitors to The Star. In order to deal with the hot and humid climate, the design applies passive ventilation with entrances on different levels and a large plaza to facilitate air flow. The passive environmental control is one of the highlights of the design, and it was the outcome of highly sophisticated analyses of daylight, solar gain, glare and so on. It has had a great effect in shaping the micro-climate within the development.

8	10
9	11

8. 星宇项目 5 000 座剧院内景
5,000-seat theatre
9. 星宇项目入口局部
Entrance to The Star
10. 一层平面图
Level 1 floor plan
11. 剖面图
Section

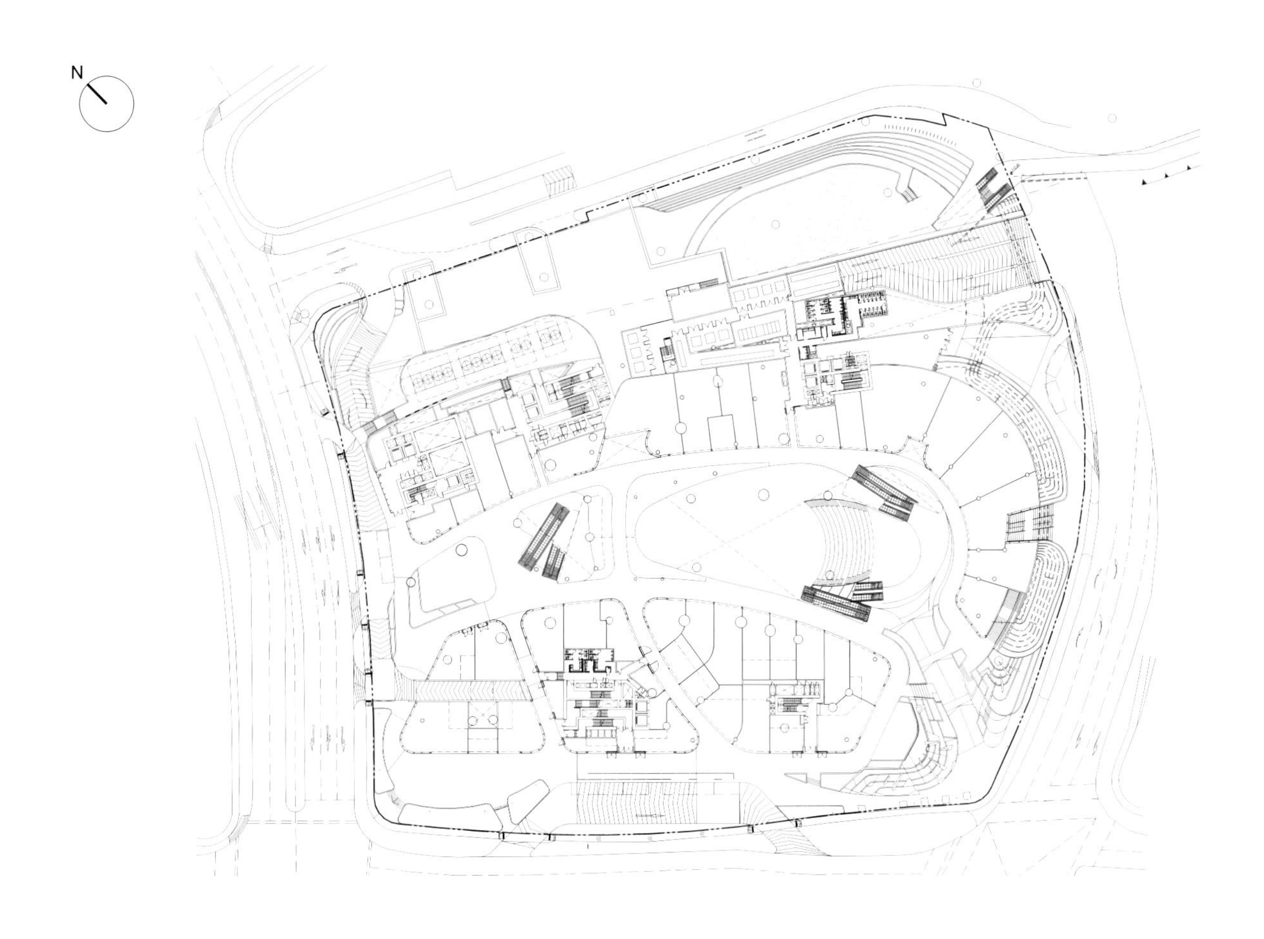
N

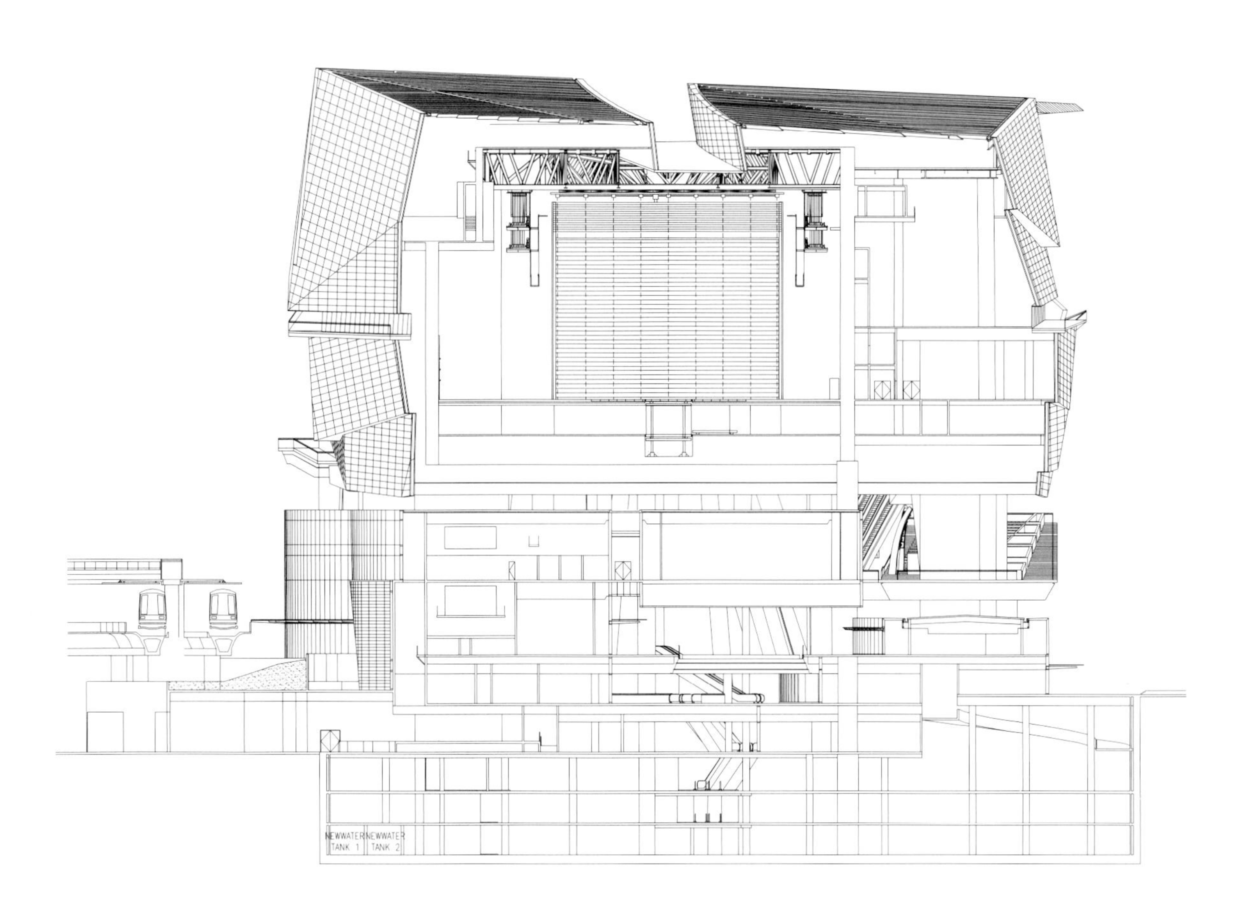
NEWWATER
TANK 1
NEWWATER
TANK 2

新加坡 Sandcrawler

Sandcrawler, Singapore

在新加坡纬壹科技城的 Fusionopolis 商业街附近，坐落着一栋外观为马蹄铁形的奇妙建筑——Sandcrawler。纬壹科技城朝气蓬勃，是汇集研发、创新和实验平台于一体的年轻态活力园区。Sandcrawler 需要符合科技城的总体规划，严格遵守建筑限高及退红线要求，并使屋顶轮廓顺应周边地形。更为挑战的是，规划中有一条城区景观轴线恰好穿过项目基地，需要设计师既要巧妙地处理建筑与城市景观的关系，同时又要确保建筑整体体量的完整性。

设计团队深入研究了周边场地及限制条件，将 Sandcrawler 大楼平面设计为“V”形，核心筒靠近凹谷底部，既顺应了地块，同时也为楼体低层创造出良好的景观视线。从立面来看，建筑由上至下层层收分，每一层都可以成为下一层的遮阴顶棚。马蹄铁形的外观增加了建筑整体的自然采光，每层都能享受绝佳的景观视野。建筑底部架空 13m，成为优质的共享空间，幽雅而郁郁葱葱的花园里种植着多种多样的热带植物，延伸入建筑底部，使建筑与城市联为一体，呈现出新加坡自然资源的丰富性和多样性，将美景还给新加坡市民和来往的旅人。

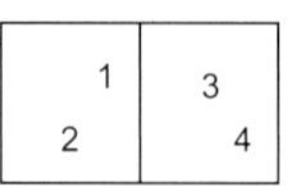

1. 新加坡城市景观
Cityscape of Singapore
2. 建筑外立面日景图
External view
3. 建筑外观
External view
4. 项目区位图
Location map

Photography © Paul Warchol

项目信息 | PROJECT INFORMATION

地　　点：新加坡纬壹
建筑面积：21 468m^2
设计时间：2009—2013 年
建成时间：2013 年
项目功能：办公楼
业　　主：Lucas Real Estate Singapore
主要设计人：Andrew BROMBERG

Location: One-north, Singapore
Gross floor area: 21,468 m^2
Design time: 2009-2013
Completion year: 2013
Sector: Office
Client: Lucas Real Estate Singapore
Design Director: Andrew BROMBERG

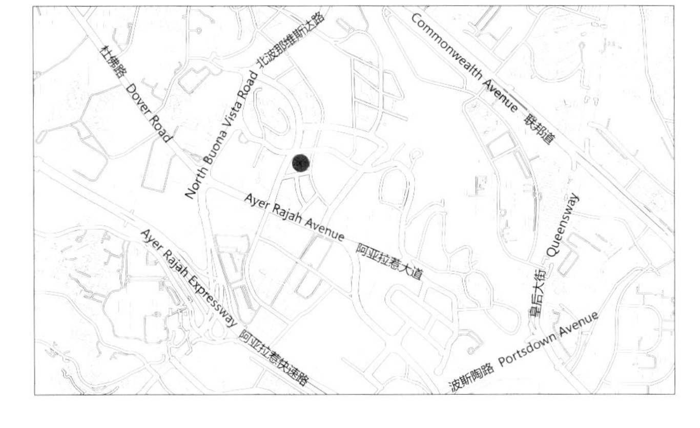

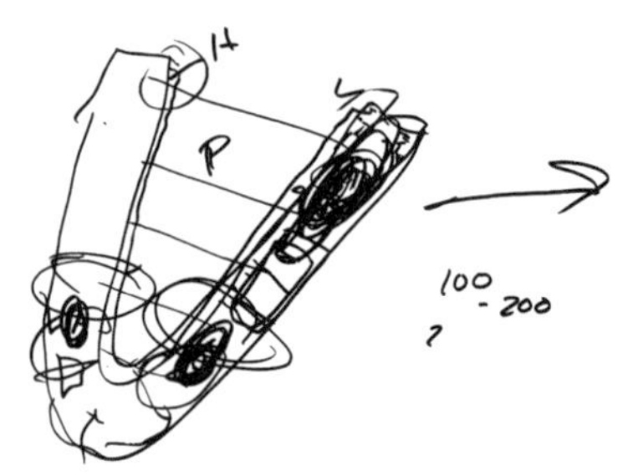

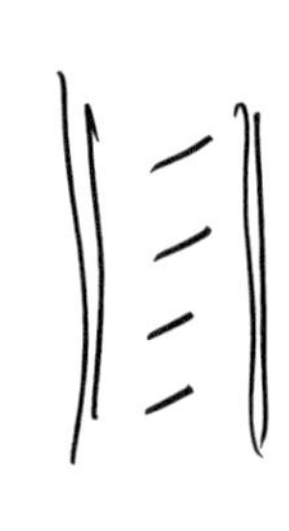

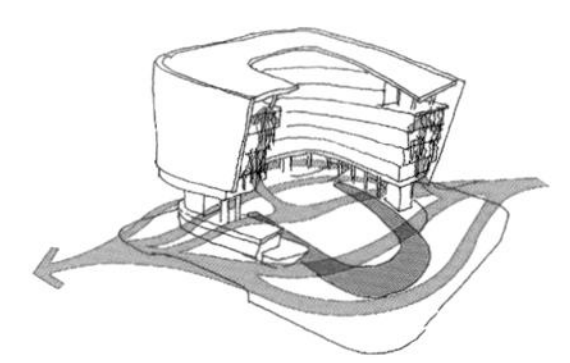

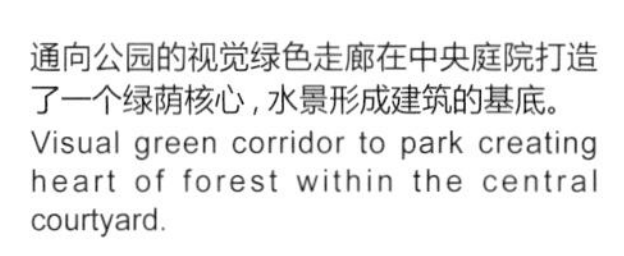

通向公园的视觉绿色走廊在中央庭院打造了一个绿荫核心，水景形成建筑的基底。
Visual green corridor to park creating heart of forest within the central courtyard.

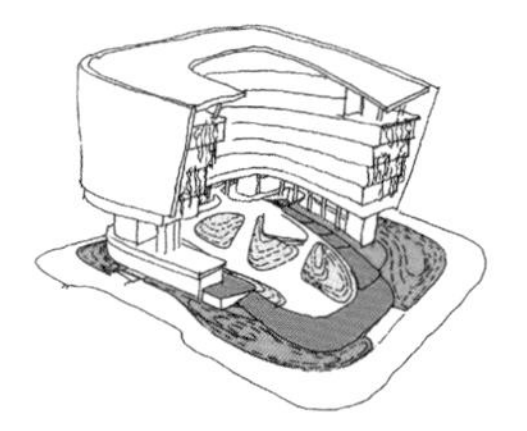

雕塑般的堆叠形成了中央庭院空间，水景增加了层次变化。
Sculptural mounding formed to create horizontal planes to central courtyard. Rill reinforces level change.

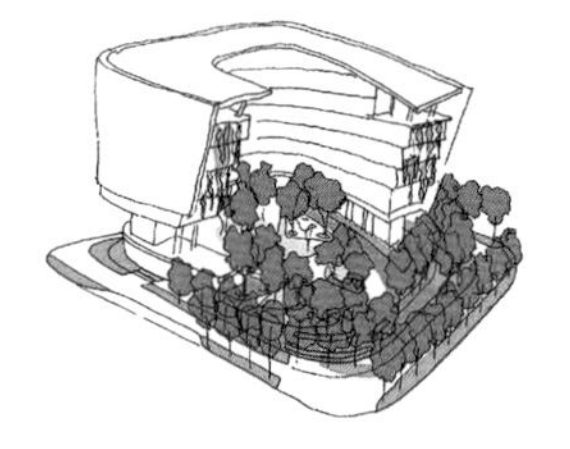

高耸的树木加强垂直绿化效果，并为场地创造封闭围合环境。
Sculptural trees to mounding, rows of large trees to perimeter to reinforce vertical green & create enclosure to the site.

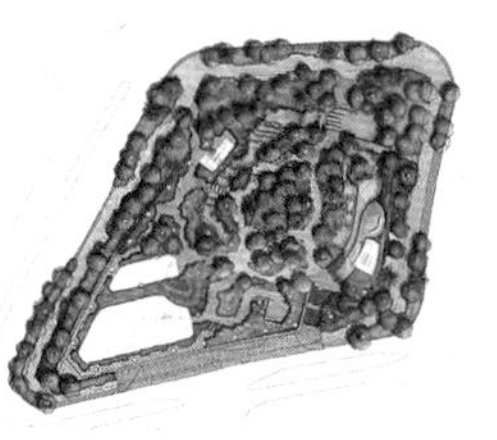

总平面图
Site plan

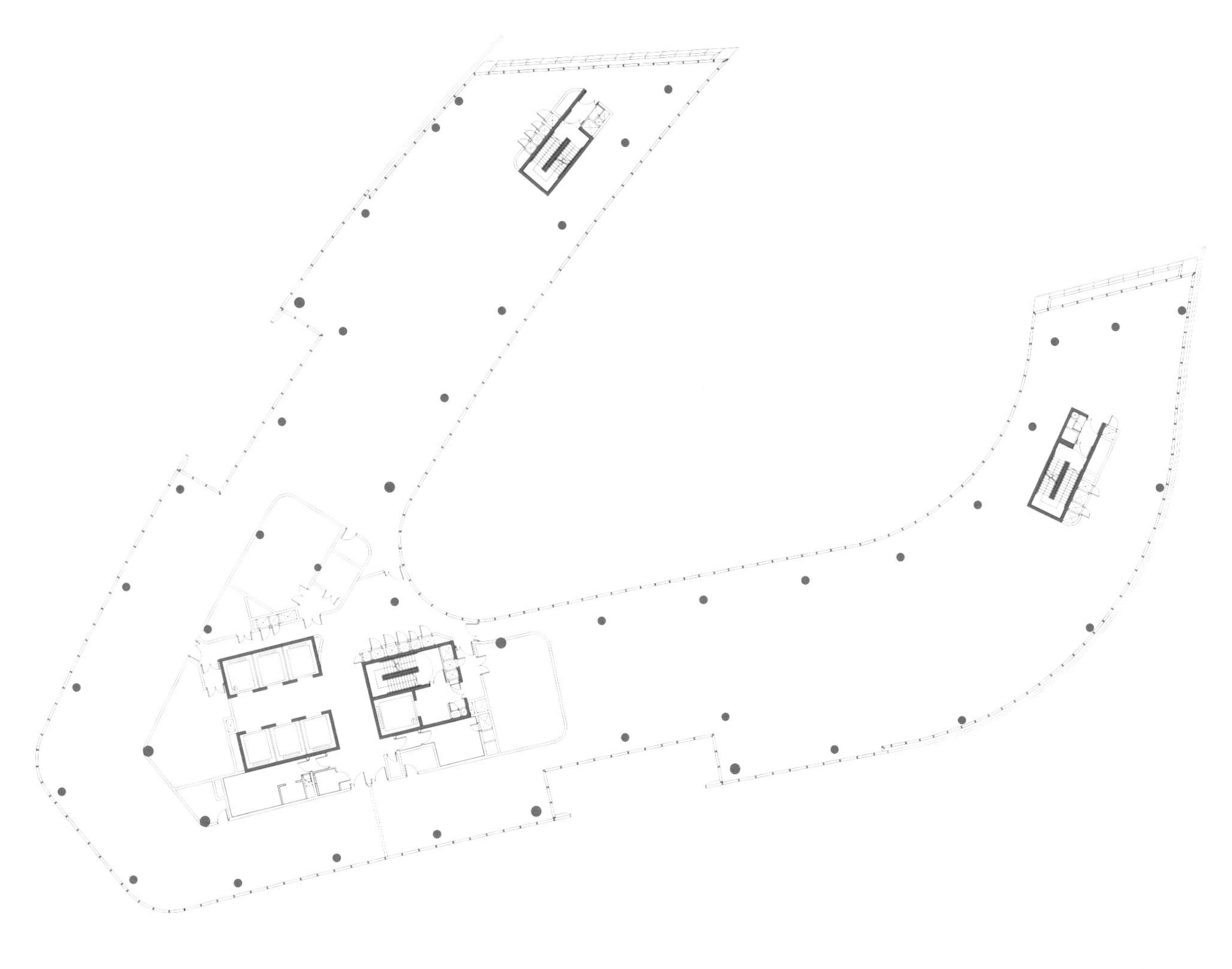

4 5 6	7

4. 围绕景观轴线的设计为低层创造了良好的景观视线
The enclosed courtyard offers excellent green view to office users
5. 手绘图
Sketch
6. 景观分析图
Landscape studies
7. 一层平面图
Level 1 floor plan

Near the Fusionopolis Commercial Street in one-north, there is a building with a V-shaped form: Sandcrawler. One-north is an energetic and active area that integrates research and development, innovation and experimentation. Sandcrawler needed to conform to the condition of its surrounding areas. It also had to abide by building height regulations and setback restrictions. Another large and difficult challenge was to accommodate a planned landscape axis that goes directly through the site, which required the designers to dedicate their skills to handle the relationship between the building and the urban landscape, while at the same time ensuring the integrity of the building itself.

The design of the nine-storey building is an extended and bent V shape with the core close to the root of the V, in response to the shape of the site and the desire to allow the landscape to flow under the building. The shape of the mass allows the lower floors to be shaded by the upper ones from direct sunshine, making each floor an excellent observation deck, and the V shape mass also facilitates natural lighting. The building is elevated from the ground by 13 meters, which creates a high-quality communal space, from which various tropical plants sprawling out into the city. The design demonstrates the variety and diversity of Singaporean culture.

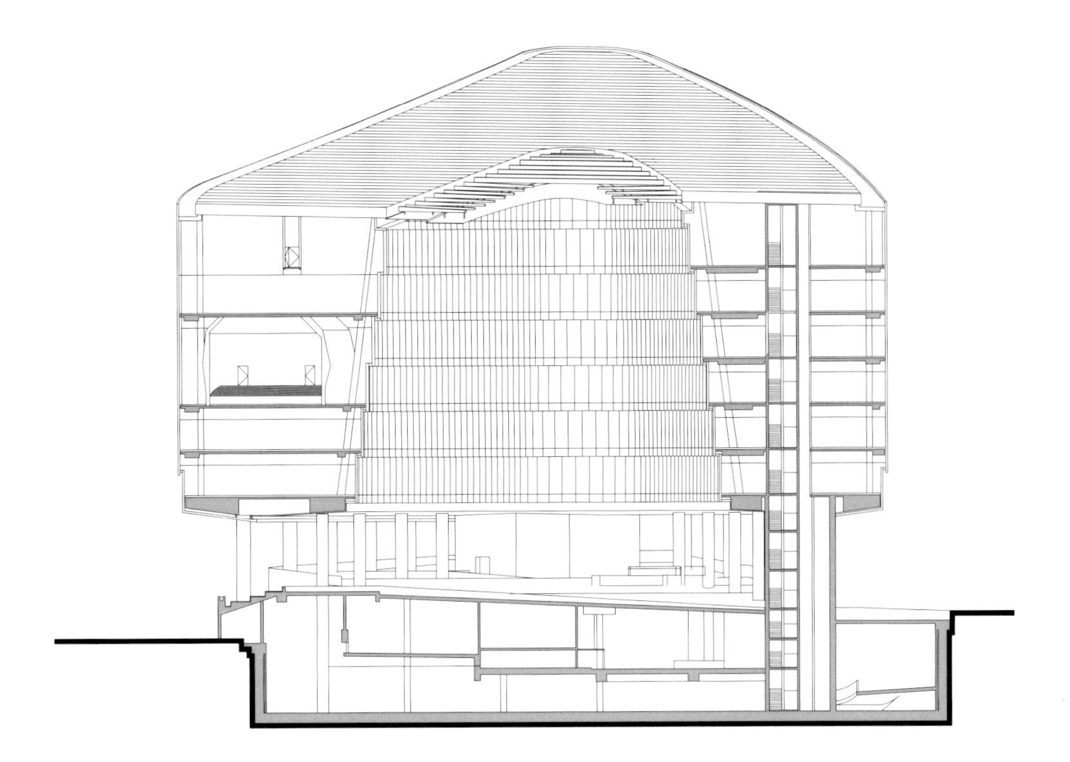

8	9
	10

8. 景观庭园
Landscaped courtyard
9. 横向剖面图
Lateral section
10. 纵向剖面图
Longitudinal section

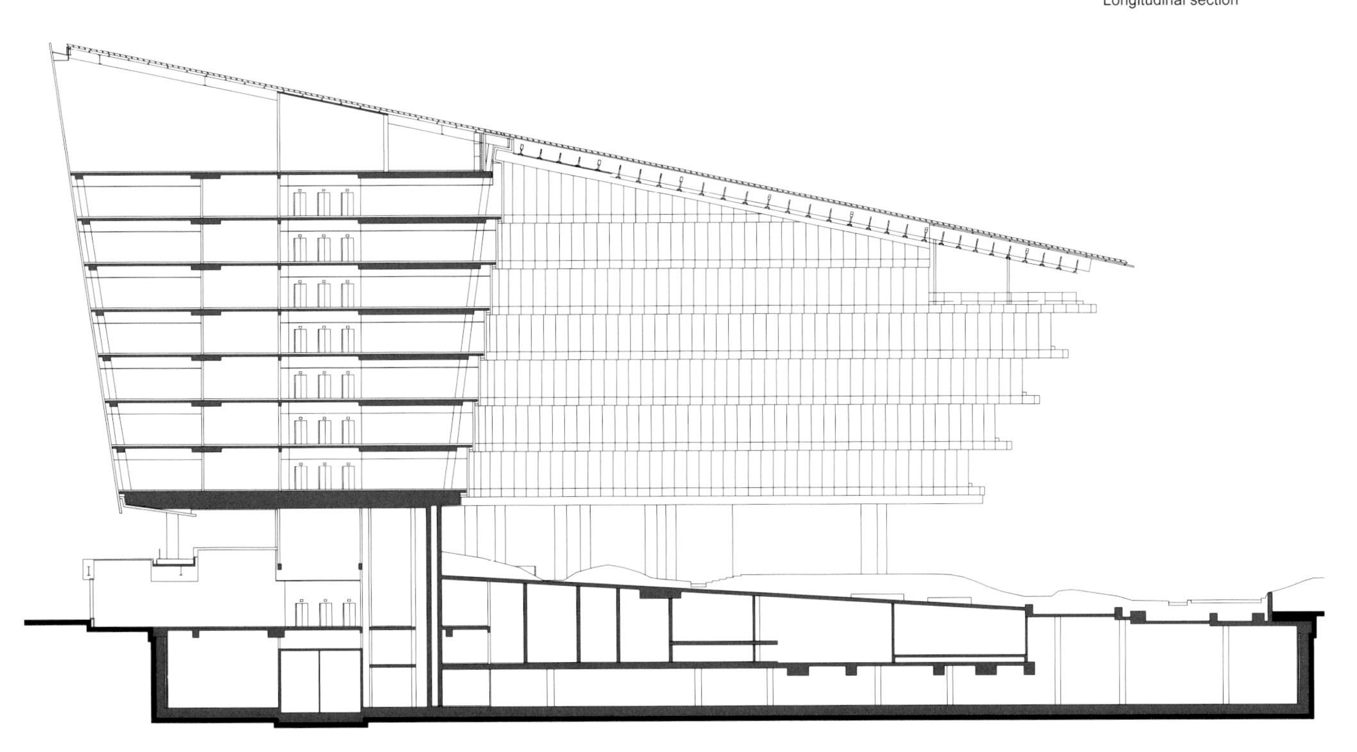

阿联酋迪拜 Ocean Heights

Ocean Heights, Dubai, UAE

在网上搜索“迪拜”“建筑”等词条，会得到诸如“疯狂的建筑奇迹”“夸张且光怪陆离的造型”之类的评价。这个在沙漠中建起的城市惯于挥金如土，每一幢建筑都在不遗余力地呐喊着自己的与众不同。迪拜码头位于蜿蜒3.34km的海岸沿线，享有独特的滨水景观，一幢幢通天的高楼用钢筋和玻璃勾勒出耀眼而鲜亮的线条。Ocean Heights 恰好矗立在这样的景观中，如何彰显自身独特的气韵成为设计师的重点思考方向。

Ocean Heights 委托方提出的要求颇具挑战，要求在动态造型的楼体内仍能保持标准化的户型系列。Aedas 以 4m 作为建筑模数，沿着高楼的一面展开并在前端开始扭转，由此让这幢 380m 高的住宅楼在外立面造型上有了韵律性的变化。令人赞叹的是，楼体的扭转仅作用于三个外立面上，而第四立面维持着持续稳定的竖直上升。在第 50 层，大楼高度已超过周边建筑，扭转外型使西南立面尺寸减小，从而降低了日光辐射。50 层及以上的公寓拥有低层公寓无法享有的绝佳景观视野，可观赏到海洋及迪拜最北端的棕榈岛。

1. 迪拜是海湾地区，与南亚次大陆隔海相望，被誉为海湾的明珠
Cityscape of Dubai
2. 远眺阿联酋迪拜 Ocean Heights
Ocean Heights stands among other towers on Business Bay
3. 建筑外观
External view

OCEAN HEIGHTS
DAMAC

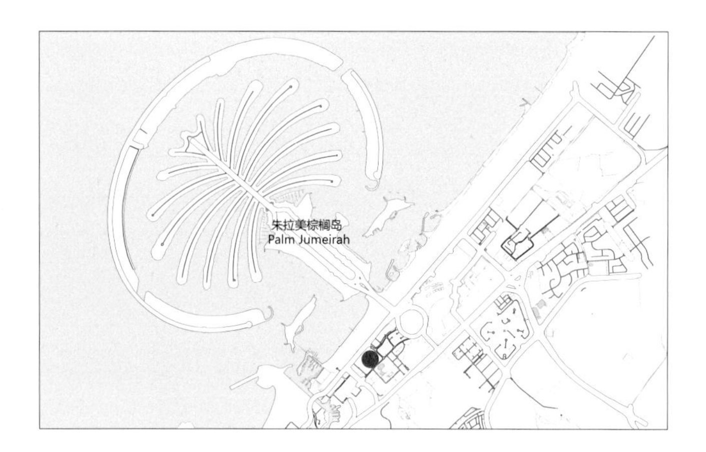

项目信息 | PROJECT INFORMATION

地　　点：阿联酋迪拜	Location: Dubai, UAE
建筑面积：79 710m^2	Gross floor area: 79,710 m^2
设计时间：2004—2010 年	Design time: 2004-2010
建成时间：2011 年	Completion year: 2011
项目功能：住宅	Sector: Residential
业　　主：DAMAC Gulf Properties LLC	Client: DAMAC Gulf Properties LLC
主要设计人：Andrew BROMBERG	Design Director: Andrew BROMBERG

If you Google "Dubai" and "Architecture," you would come across descriptions of the city as an "architectural miracle" with an "exaggerated and peculiar form." The desert city built is certainly extravagant, and every single building seems to flaunt its distinctiveness. The Dubai Marina area runs along 3.34km of coastline and it possesses unique waterfront views. Its skyscrapers compose a marvelous scene with concrete and glass. This is the landscape in which Ocean Heights is located. The challenge was how to express its own character in place where each building competes for attention with the spectacular scenery.

The client's requirement called for the building to have a dynamic form with a consistent plan for all floors. That was not obviously possible with the building's feature twist. Aedas' way of dealing with this was to establish a system of standard four-meter wide modules. Their sides stack up all the way down the building, with only the front part of the module changing to produce the exterior twist. The building's twist starts from the base on three of the four sides. The fourth side remains constantly orthogonal all the way up. By level 50, the building is taller than its neighbors and the area of southwest elevation is reduced so as to minimise solar gain. Equally importantly, on level 50 and above contains apartments with views toward the ocean and Dubai's most northerly Isle of Palms.

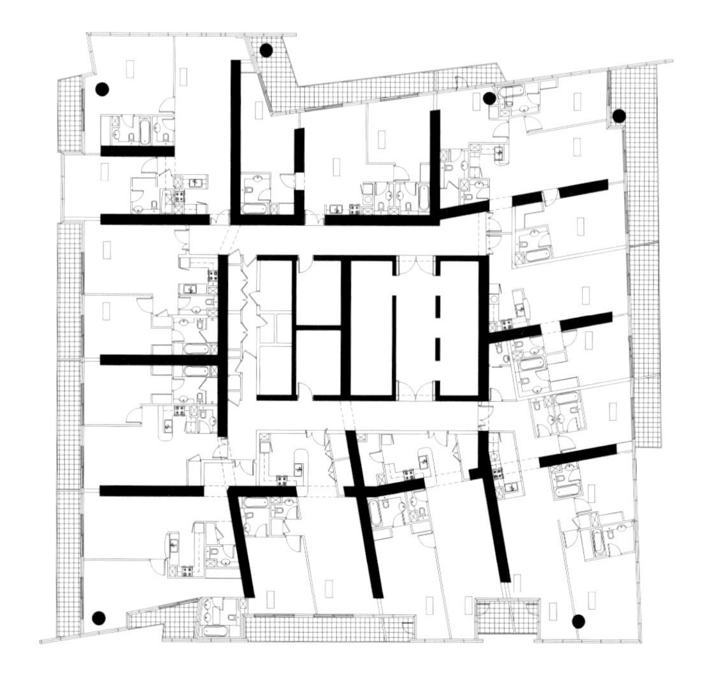

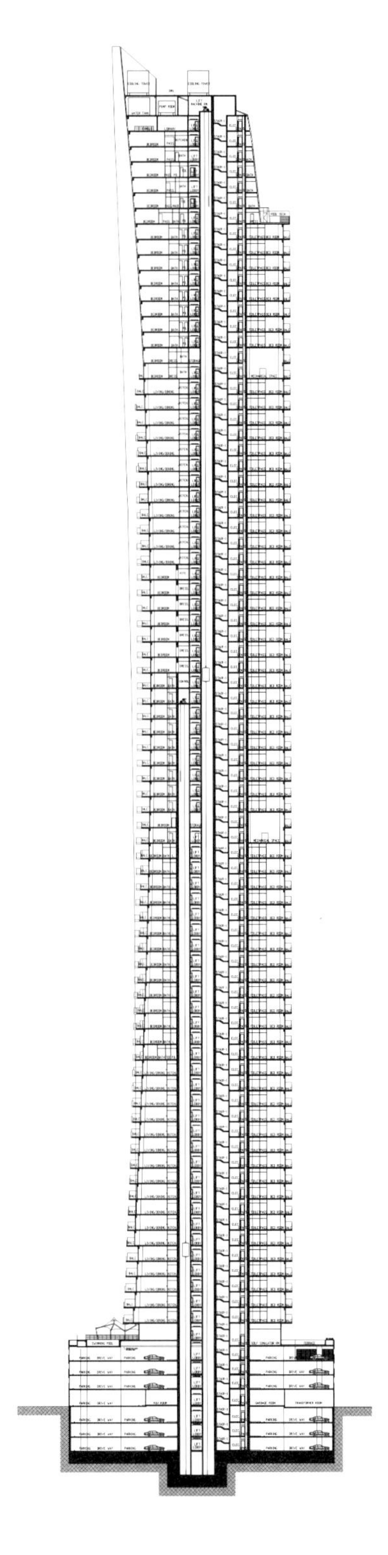

4 5 6	7 8

4. 项目区位图
Location map
5. 外立面细部
Facade details
6. 标准层平面图
Typical floor plan
7. 剖面图
Section
8. 建筑拥有绝佳景观视野，可观赏到海洋及迪拜最北端的棕榈岛
The building commands an excellent view of the ocean and the Isle of Palms

阿联酋迪拜地铁站

Dubai Metro, Dubai, UAE

高速发展的时代洪流中，来往穿行的地铁成为城市文明和社会生活最直观而真实的缩影，从入口到出口，人们以最迅捷的方式切身感受着城市的脉搏。Aedas 已成功完成了多伦多、吉隆坡、德里、新加坡、中国香港等重要城市轨交系统的设计，而其中一个项目是阿联酋迪拜地铁站。作为海湾国家首个城市地铁交通网络系统的开发者，迪拜地铁管理局在全长 74km 的轨道交通网设计中向 Aedas 提出了极具挑战的要求：独特的外形、地标性的高辨识度以及统一的整体风格。

设计团队沉浸在旧时迪拜采珠村的记忆提炼中，小绳系腰、没水取珠的昔日繁华深深打动了每位设计师，也激发了他们的设计灵感：以珠母贝为原型，延续其粗粝外壳与细滑内表面的强烈反差，站台以蚌壳轻覆沿线轨道，点缀并串联起迪拜的轨道交通。为体现风格的统一，每个车站都采用双层车站的设计策略，一层设置站台，另一层设置售票处和零售店。一颗颗晶亮的明珠，贯联了快速交通网络，也贯联了这座城市的过去和未来。

1. 迪拜城市景观
Dubai city
2. 地铁站人视外观
Perspective view of a station
3. 地铁站内部垂直循环交通空间
Vertical circulation

项目信息 | PROJECT INFORMATION

地　　点：阿联酋迪拜
项目规模：地铁线路总长 74 km，共有 45 个地铁站点
设计时间：2006 年
建成时间：2009 年（I 期）
项目功能：基础设施
业　　主：迪拜道路及交通局

Location: Dubai, UAE
Project scale: Total of 74km of lines and 45 stations
Design time: 2006
Completion year: 2009 (Phase 1)
Sector: Infrastructure
Client: Roads and Transport Authority of Dubai

In an era of high-speed development, rapid transit networks have become the most tangible and realistic microcosm of urban civilisation and social life. From entrance to exit, people feel the quick-beating pulse of the city. Aedas has designed very successful mass transit rail projects in Toronto, Kuala Lumpur, Delhi, Hong Kong, Singapore and elsewhere. One of the projects was the Dubai Metro in the United Arab Emirates. As the developer of the first urban metro network in the Gulf countries, the Roads and Transport Authority of Dubai presented Aedas with challenging requirements for a 74km long automated passenger transit project. They had to work well, but they also had to be memorably eye-catching and coherent in style.

The design team developed their most fruitful set of ideas when they pondered the history and tradition of old Dubai and its history as a pearl-diving village. They drew inspiration from oyster shells and explored how they might be represented in the design. In order to maintain a coherent style of design across the system, each of the stations is based on two levels, with platforms on one and ticket office and retail on the other. The stations are like jewels in a necklace of express transportation, reflecting the city's history and future.

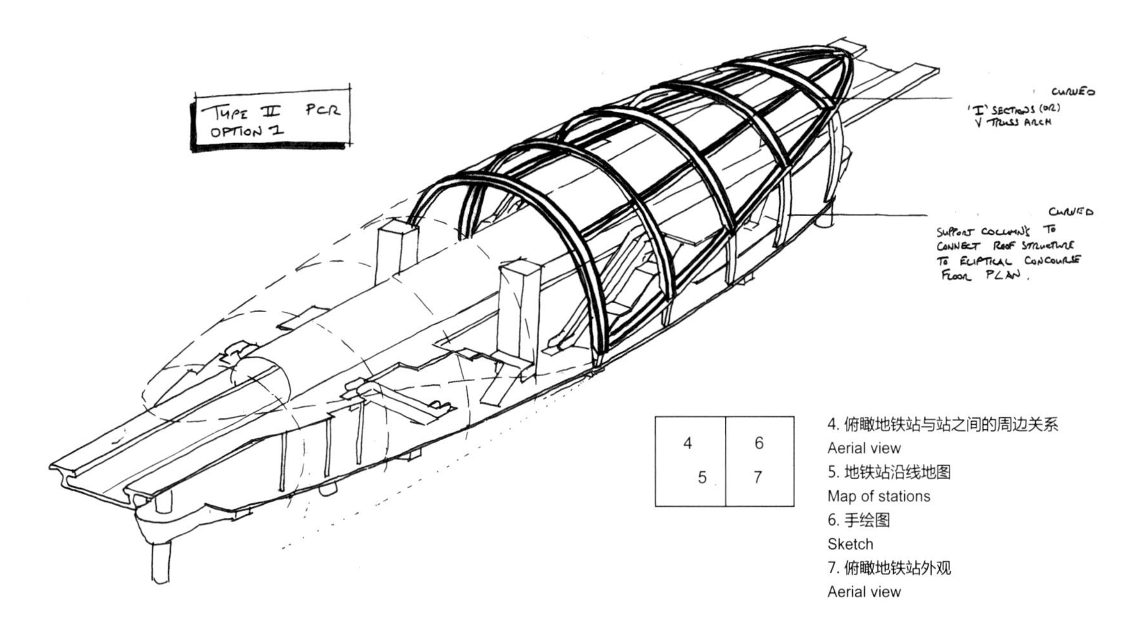

4	6
5	7

4. 俯瞰地铁站与站之间的周边关系
Aerial view
5. 地铁站沿线地图
Map of stations
6. 手绘图
Sketch
7. 俯瞰地铁站外观
Aerial view

إلى جبل علي
42 To Jebel Ali
Alokozay
VANILLA TEA
انتبه للمسافة الفاصلة
Out
خروج

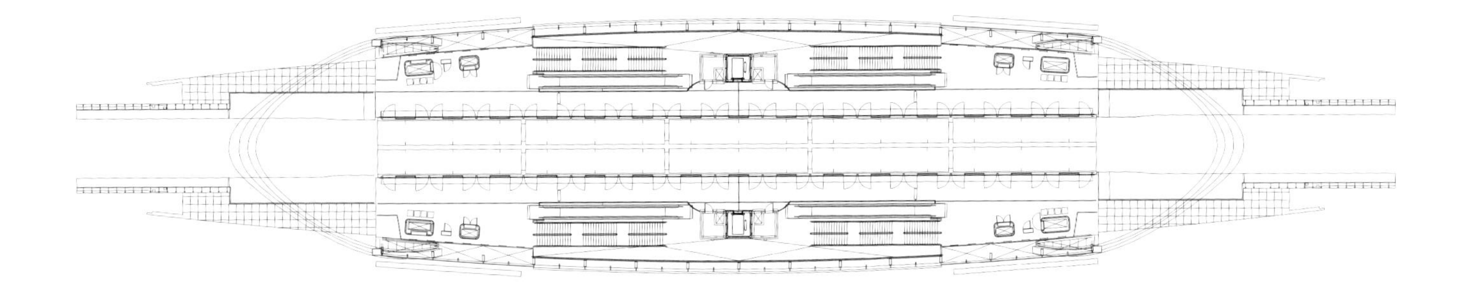

8	10
9	11 12

8. 地铁站内部垂直循环交通空间
Vertical circulation
9. 候车站台
Platform
10. 站台层平面图
Platform floor plan
11.12. 剖面图
Section

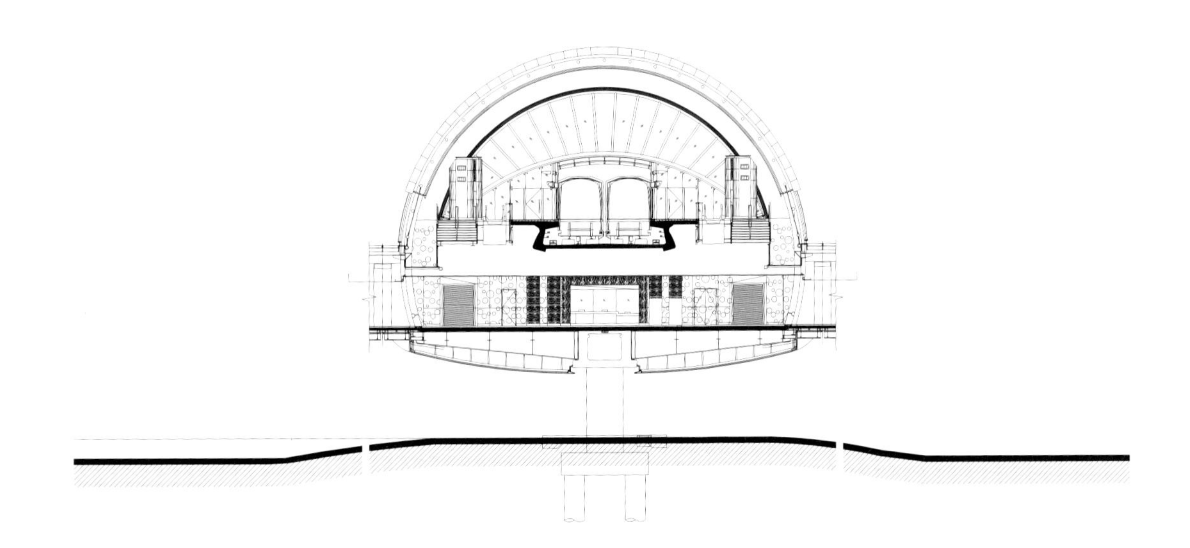

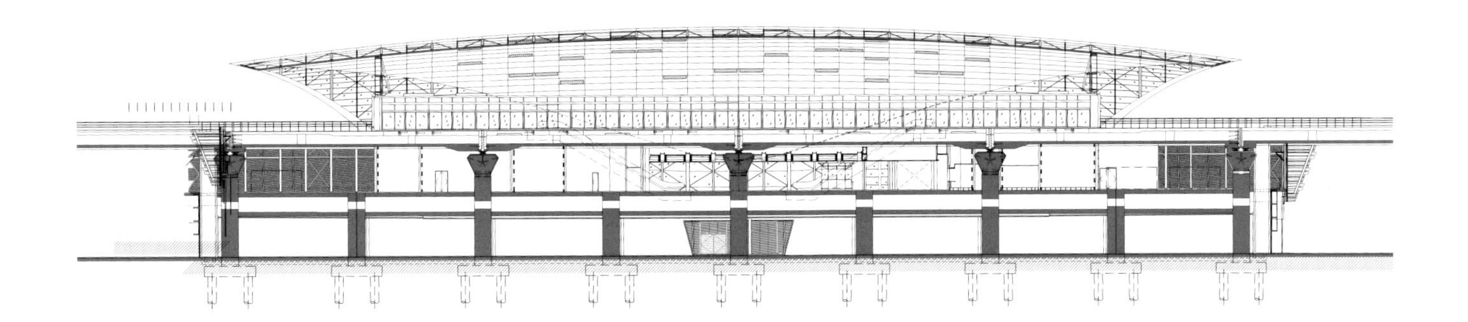

英国唐卡斯特 Cast 剧院

Cast, Doncaster, UK

Cast 剧院项目始于英国唐卡斯特文化社区 MUSE Waterdale 的总体规划。唐卡斯特，有着独特的工业和商业传统，也有着新兴的创新产业；当地矮小茂盛的植物沼泽地、石楠花、风刮树、暴露的岩层，都彰显出约克郡独有的自然风貌，这些都为设计提供了灵感素材。

Cast 剧院被设计为一个“文化工具箱”，具有很高的灵活性来适应当前及未来的市场发展需求。这个“工具箱”拥有 600 席的歌剧院，200~400 席的大平层演奏厅，舞蹈工作室，戏剧工作室及 3 个灵活布局的教育空间。

Cast 剧院的建筑柱网布局顺应西北方向的法院和警察局，引导出从城镇中心到东南方向的步行线路，为新建广场创造了瞩目的临街立面。

Cast 剧院是唐卡斯特文化社区的公共活动中心，可持续发展理念渗透在所有建筑元素中。观众席、照明、玻璃外壳是 Cast 设计的三大亮点。观众席的暖红金色调营造出温暖、戏剧性和包裹性的氛围。剧院照明是基于突显钢索（唐卡斯特的工业遗产之一）的横截面而设计的。玻璃外壳回应着周边的新地方议会办公室、新景观广场和公共图书馆，建筑通透而有存在感和亲民感，夜间格外引人注目，而这对于剧院设计至关重要。材质上，使用了当地石灰岩以确保颜色和质感体现当地自然环境的特色；再造石和铜的工业感体现出唐卡斯特的历史气质。

1	3
2	4

1. 唐卡斯特独特的工商业传统和新兴创新产业为设计提供了灵感素材
Doncaster has a unique industrial and commercial tradition as well as an emerging innovative industry, which provide great inspiration to the design
2. Cast 剧院正面人视
External view
3. 表演厅观众席
Auditorium
4. 项目区位图
Location map

Photography © Philip Vile

项目信息 | PROJECT INFORMATION

地　　点：英国唐卡斯特	Location: Doncaster, UK
建筑面积：5 840m^2	Gross floor area: 5,840 m^2
设计时间：2008—2013 年	Design time: 2008-2013
建成时间：2013 年	Completion year: 2013
项目功能：文化与休闲	Sector: Arts and Leisure
业　　主：MUSE Developments	Client: MUSE Developments
主要设计人：Julian MIDDLETON	Director: Julian MIDDLETON

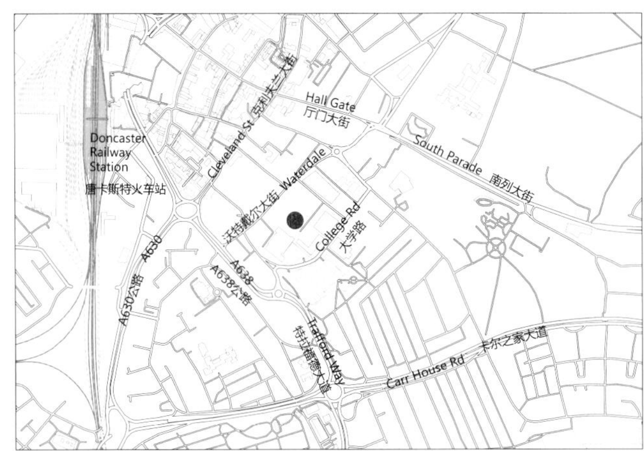

Photography © Philip Vile

Cast is designed as a key element of the MUSE Waterdale master plan for a civic and cultural quarter in the heart of Doncaster, a city in Yorkshire. Doncaster has a unique industrial and commercial heritage as well as current innovative industries. Yorkshire's unique natural landscape, which features moors with low scrubby flora such as heathers and windswept trees, as well as exposed rock formations, has influenced the design.

Cast is designed as a creative toolbox, a flexible, responsive contemporary venue for current and future markets. It includes a 600-seat fixed format theatre; a flexible, flat floor pesformance area

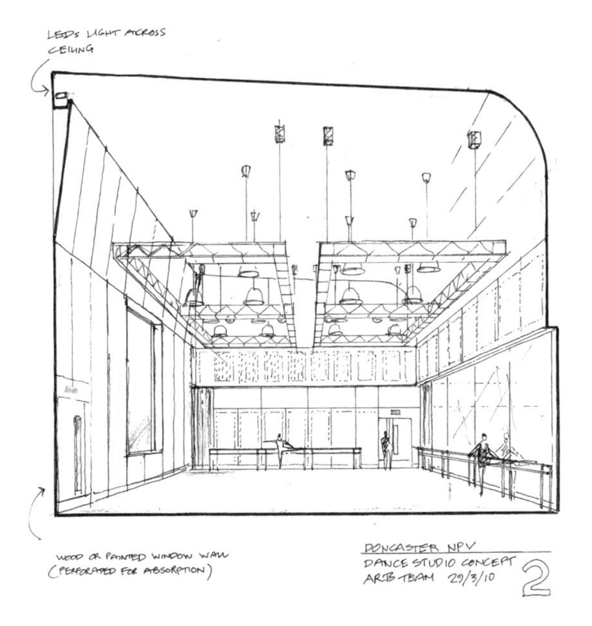

5	7 8
6	9

5. 一层公共空间
Communal space on ground floor
6. 舞蹈工作室
Dance studio
7. 观众席手绘图
Sketch of auditorium
8. 舞蹈工作室手绘图
Sketch of dancing studio
9. 俯瞰表演厅观众席
Auditorium

for 200 to 400 capacity; a suite of education spaces including dance studio, drama studio, and three flexible teaching spaces.

The building is on grid in alignment with the court building and police station which are located to the northwest. The design has successfully reinforced the pedestrian route from the town center to the southeast and delivered a strong street frontage onto the newly built square.

Cast is a public activity center for the civic and cultural quarter of Doncaster, and the concept of sustainable development permeates all of its architectural elements. The auditorium, lighting and glazed facade are the three key aspects of the design. The warm reds and golds of the auditorium create a warm, dramatic and enveloping atmosphere. The light designs are based on the cross section of steel ropes — part of Doncaster's industrial heritage. The glazed facade facing the front of the building is a design response to the new offices of the local council, as well as the new landscaped square and the public library, allowing the building to be transparent, strong and civic-minded. It is brightly lit and inviting in the evening, which is a highly important element of theatre design. Cast uses local limestone to ensure the colour and texture reflect the regional surroundings. The industrial look and feel of the reconstituted stone and copper reflects the historical character of Doncaster.

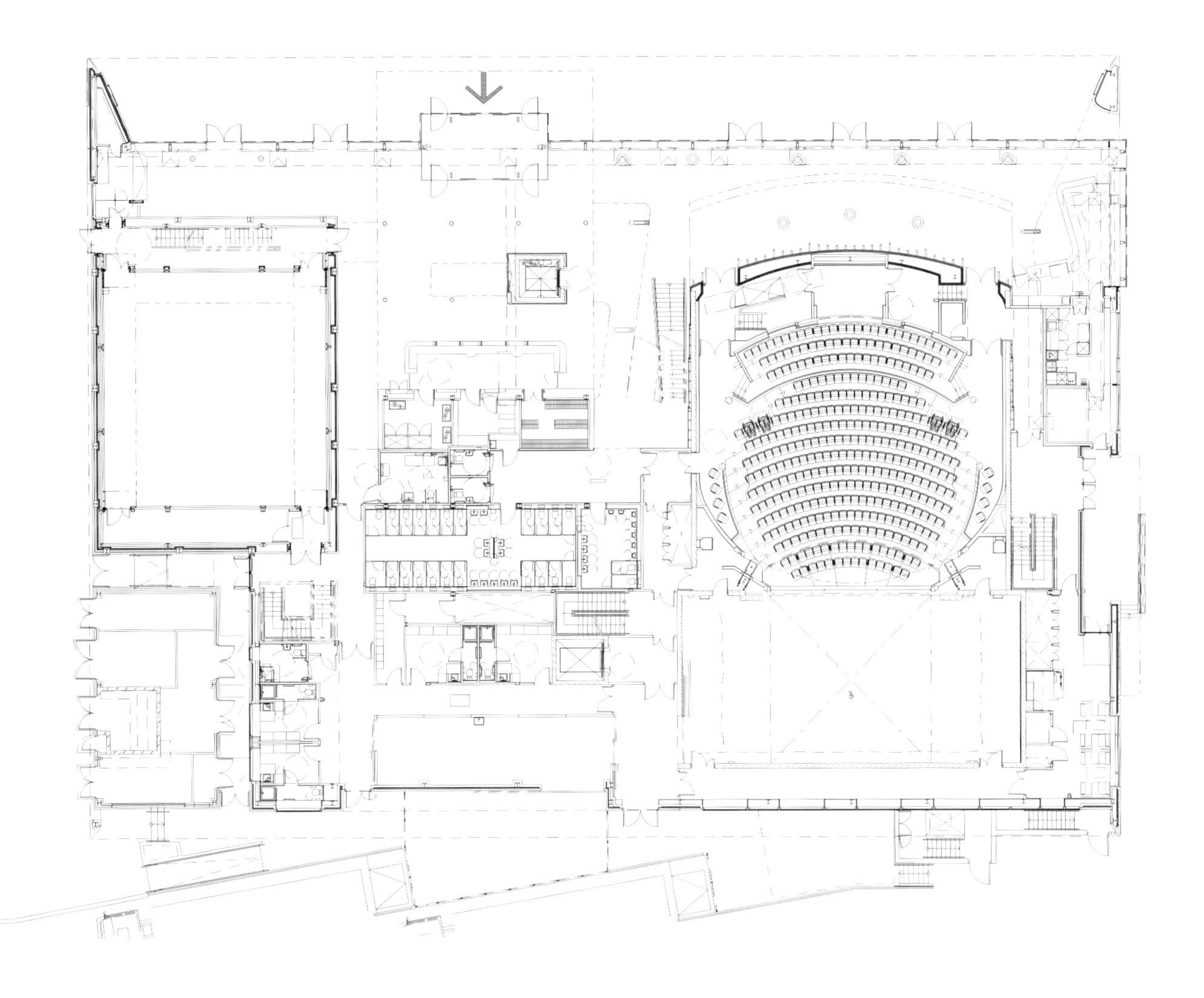

10	11
	12

10. 小剧场
Studio theatre
11. 一层平面图
Level 1 floor plan
12. 剖面图
Section

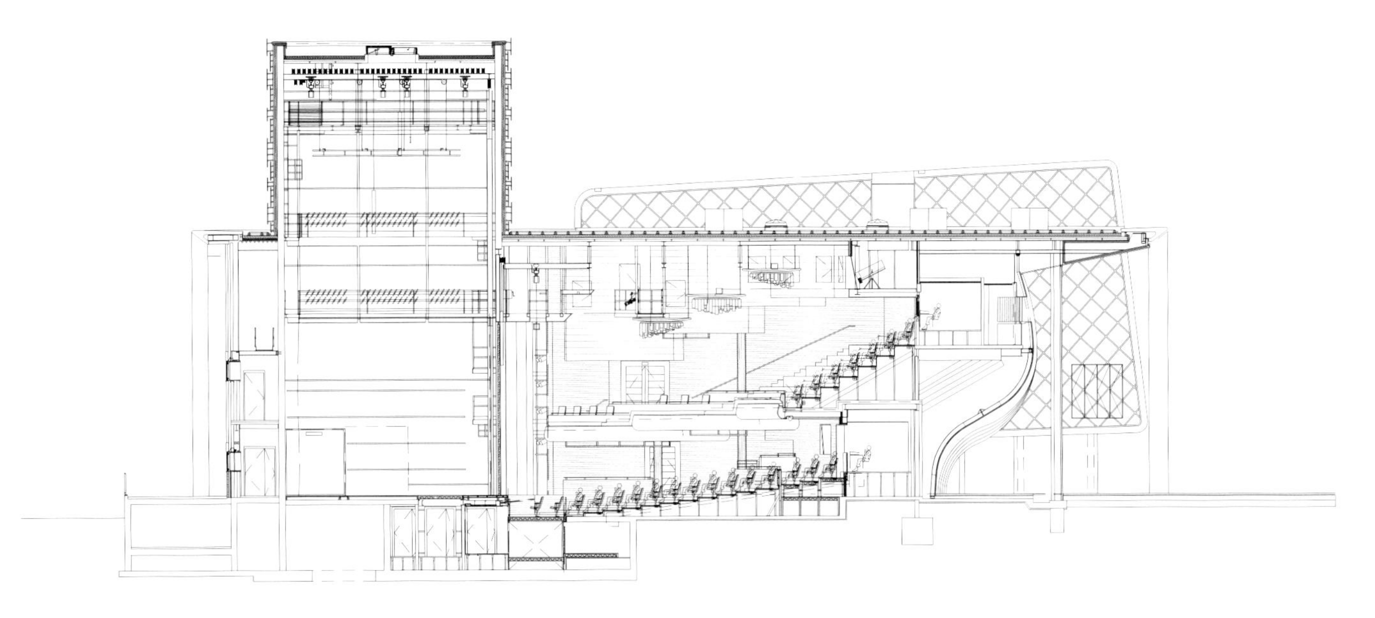

附录 1 APPENDIX 1
AWARDS
荣获奖项

获奖信息统计时间截止至：2018 年 8 月
As of August 2018

NO.01 大连恒隆广场 Olympia 66, Dalian

· 2018 年 ICSC 中国购物中心 & 零售商大奖
——设计和开发新发展项目金奖
· ICSC China Shopping Center & Retailer Awards 2018
—Gold Award, Design and Development, New Developments

· 2018 年全球 RLI 大奖
——RLI 国际购物中心项目优异奖
· Global RLI Awards 2018
—Highly Commended, RLI International Shopping Center

· 2017 年 MIPIM ASIA 大奖
——最佳零售发展项目银奖
· MIPIM Asia Awards 2017
—Silver, Best Retail Development

· 2017 年 AAP 建筑大奖
——建筑设计商业建筑荣誉奖
· AAP Architecture Prize 2017
—Honorable Mention, Architectural Design Commercial Architecture

· 2017 年亚太购物中心大奖
——设计和开发新发展项目金奖
——设计和开发新发展项目可持续发展设计奖
· Asia Pacific Shopping Center Awards 2017
—Gold Award, Design and Development, New Developments
—Sustainability Design Award, Design and Development, New Developments

· 2016 IDA 国际设计奖
——专业建筑—新商业建筑项目荣誉奖
· IDA International Design Awards 2016
—Honorable Mention, Professional Architecture, New Commercial Building

· 2016—2017 年 A' 设计奖
——建筑，楼宇及结构设计金奖
· A' Design Awards 2016-2017
—Golden Winner, Architecture, Building and Structure Design

· 2016 年 Cityscape 新兴市场建筑大奖
—— 已建零售项目优异奖
· Cityscape Awards for Architecture in Emerging Markets 2016
—Highly Commended, Built Retail Project Award

· 2013 年 Cityscape 新兴市场建筑大奖
—— 未来零售项目优胜奖
· Cityscape Awards for Architecture in Emerging Markets 2013
—Winner, Future Retail Project Award

· 2011 年国际房地产大奖
—— 最佳零售建筑优胜奖
· International Property Awards 2011
—Winner, Best Retail Architecture

· 2011 年 MIPIM ASIA 大奖
——最佳中国未来建设项目铜奖
· MIPIM Asia Awards 2011
—Bronze Winner, Best Chinese Futura Projects

· 2011 年亚太商业房地产大奖
——亚太最佳零售建筑优胜奖
——中国最佳零售建筑 5 星奖
· Asia Pacific Commercial Property Awards 2011
—Winner, Best Retail Architecture, Asia Pacific
—5-Star, Best Retail Architecture, China

NO.02 北京大望京综合开发项目 Da Wang Jing Mixed-use Development, Beijing

· 2013 年亚太房地产大奖
——中国最佳高层建筑 5 星奖
· Asia Pacific Property Awards 2013
—5-Star, Best High Rise Architecture, China

NO.03 北京大兴 3 及 4 地块项目 Daxing Plots 3 and 4, Beijing

· 2017 年 AAP 建筑大奖
——建筑设计综合建筑优胜奖
· AAP Architecture Prize 2017
—Winner, Architectural Design Mixed-Use Architecture

· 2013 年国际房地产大奖
——最佳综合建筑优胜奖
· International Property Awards 2013
—Winner, International Award, Best Mixed-Use Architecture

· 2013 年亚太房地产大奖
——亚太最佳综合建筑优胜奖
——中国最佳综合建筑 5 星奖
· Asia Pacific Property Awards 2013
—Winner, Best Mixed-Use Architecture, Asia Pacific
—5-Star, Best Mixed-Use Architecture, China

· 2012 年 MIPIM ASIA 大奖
——最佳中国未来大型建设项目金奖
· MIPIM Asia Awards 2012
—Gold Winner, Best Chinese Futura Mega Project

NO.04 北京新浪总部大楼 Sina Plaza, Beijing

· 2016 IDA 国际设计奖
——专业建筑—新商业建筑项目荣誉奖
· IDA International Design Awards 2016
—Honorable Mention, Professional Architecture, New Commercial Building

· 2016—2017 年 A' 设计奖
——建筑，楼宇及结构设计金奖
· A' Design Awards 2016-2017
—Golden Winner, Architecture, Building and Structure Design

· 2016 年 MIPIM ASIA 大奖
—— 最佳办公楼及商业发展项目金奖
· MIPIM Asia Awards 2016
—Gold, Best Office & Business Development

· 2016 年亚洲建 + 设大奖
—— 商业，零售及办公楼〈专业〉建筑优异奖
· A&D Trophy Awards 2016
—Certificate of Excellence, Architecture (Professional) Commercial, Retail or Office

· 2016 年亚太房地产大奖
——中国办公楼建筑优异奖
· Asia Pacific Property Awards 2016
—Highly Commended, Office Architecture, China

· 2015 年 MIPIM ASIA 大奖
——最佳中国未来建设项目银奖
· MIPIM Asia Awards 2015
—Silver Winner, Best Chinese Futura Project

NO.05 北京北苑北辰综合体
North Star Mixed-use Development, Beijing

· 2012 年《透视》杂志大奖
——综合用途〈专业〉建筑优异奖
· Perspective Awards 2012
—Certificate of Excellence, Architecture (Professional) Mixed-Use

· 2011 全国人居经典规划设计方案竞赛
——最佳设计方案金奖
· China's Outstanding Architectural Design & Planning Award 2011
—Gold Award, Best Design Award

· 2011 年上海建筑学会商用建筑大奖
——创新奖
· Shanghai Commercial Real Estate Professional Committee of ASSC 2011
—Innovation Award

· 2010 年亚太商业房地产大奖
——亚太综合建筑优胜奖
——中国综合建筑 5 星奖
· Asia Pacific Commercial Property Awards 2010
—Winner, Mixed-use Architecture Award, Asia Pacific
—5-Star, Mixed-use Architecture Award, China

· 2010 年 Cityscape 新兴市场建筑大奖
——已建商业及综合发展项目优异奖
· Cityscape Awards for Architecture in Emerging Markets 2010
—Highly Commended, Built Commercial / Mixed-use Projects

NO.06 青岛金茂湾购物中心
Jinmao Harbour Shopping Center, Qingdao

· 2015 年 Cityscape 新兴市场建筑大奖
——未来零售项目优胜奖
· Cityscape Awards for Architecture in Emerging Markets 2015
—Winner, Future Retail Project Award

NO.07 成都恒大广场
Evergrande Plaza, Chengdu

· 2016 年国际房地产大奖
——国际最佳酒店建筑优胜奖
· International Property Awards 2016
—Winner, Best International Hotel Architecture

· 2016 年亚太房地产大奖
——亚太最佳酒店建筑优胜奖
——中国最佳酒店建筑 5 星奖
——中国商业高层建筑优异奖
· Asia Pacific Property Awards 2016
—Winner, Best Hotel Architecture, Asia Pacific
—5-Star, Best Hotel Architecture, China
—Highly Commended, Commercial High-rise Architecture, China

NO.08 重庆新华书店集团公司解放碑时尚文化城
Chongqing Xinhua Bookstore Group Jiefangbei Book City Mixed-use Project, Chongqing

· 2018 年亚太房地产大奖
——中国综合建筑优胜奖
· Asia Pacific Property Awards 2018
—Winner, Mixed-use Architecture, China

· 2017 年 AAP 建筑大奖
——建筑设计高层建筑优胜奖
· AAP Architecture Prize 2017
—Winner, Architectural Design Tall Buildings

· 2017 年香港建筑师学会两岸四地建筑设计大奖
——未兴建项目：建筑方案设计组别卓越奖
· HKIA Cross-strait Architectural Design Awards 2017
—Nominated Award, Un-built Projects: Architectural Design Scheme

· 2016—2017 年 A' 设计奖
——建筑，楼宇及结构设计白金奖
· A' Design Awards 2016-2017
—Platinum Winner, Architecture, Building and Structure Design

· 2016 年 MIPIM ASIA 大奖
——最佳中国未来大型建设项目银奖
· MIPIM Asia Awards 2016
—Silver, Best Chinese Futura Mega Project

NO.09 武汉恒隆广场
Heartland 66, Wuhan

· 2016—2017 年 A' 设计奖
——建筑，楼宇及结构设计金奖
· A' Design Awards 2016-2017
—Golden Winner, Architecture, Building and Structure Design

· 2015 年 Cityscape 新兴市场建筑大奖
——未来综合发展项目优胜奖
· Cityscape Awards for Architecture in Emerging Markets 2015
—Winner, Future Mixed-use Project Award

· 2015 年亚太房地产大奖
——中国综合建筑优异奖
· Asia Pacific Property Awards 2015
—Highly Commended, Mixed-use Architecture, China

NO.10 上海星荟中心
Shanghai Landmark Center, Shanghai

· 2017 年 IDA 国际设计奖
——建筑类—新商业建筑项目铜奖
· IDA International Design Awards 2017
—Bronze, Architecture Category – New Commercial Building

· 2017—2018 年 A' 设计奖
——建筑，楼宇及结构设计银奖
· A' Design Awards 2017-2018
—Silver Winner, Architecture, Building and Structure Design

· 2017 年 AAP 建筑大奖
——建筑设计商业建筑荣誉奖
· AAP Architecture Prize 2017
—Honorable Mention, Architectural Design Commercial Architecture

· 2017 年亚洲建 + 设大奖
——商业，零售及办公楼〈专业〉建筑优异奖
· A&D Trophy Awards 2017
—Certificate of Excellence, Architecture (Professional) Commercial, Retail or Office

· 2016—2017 年香港建筑师学会年奖
——境外商业建筑优异奖
· The Hong Kong Institute of Architects Annual Awards 2016-2017
—Merit Award, Commercial Building Outside Hong Kong

NO.11 上海龙湖虹桥项目
Longfor Hongqiao Mixed-use Project, Shanghai

· 2013 年亚太房地产大奖
——中国零售建筑优异奖
· Asia Pacific Property Awards 2013
—Highly Commended, Retail Architecture, China

NO.12 上海虹桥世界中心
Hongqiao World Center, Shanghai

· 2015 年亚太房地产大奖
——中国综合建筑优异奖
· Asia Pacific Property Awards 2015
—Highly Commended, Mixed-use Architecture, China

NO.13 无锡恒隆广场
Center 66, Wuxi

· 2016 年优质建筑大奖
——境外非住宅建筑项目优异奖
· Quality Building Award 2016
—Merit Award, Building Outside Hong Kong (Non-Residential)

· 2015 年亚洲建 + 设大奖
——商业，零售及办公楼〈专业〉建筑优异奖
· A&D Trophy Awards 2015
—Certificate of Excellence, Architecture (Professional) Commercial, Retail or Office

· 2015 年全球 RLI 大奖
——RLI 国际购物中心项目优异奖
· Global RLI Awards 2015
—Highly Commended, RLI International Shopping Center

· 2015 年香港建筑师学会两岸四地建筑设计大奖
——商场 / 步行街组别卓越奖
· HKIA Cross-strait Architectural Design Awards 2015
—Nominated Award, Commercial: Shopping

· 2014 年 MIPIM ASIA 大奖
——最佳综合发展项目银奖
· MIPIM Asia Awards 2014
—Silver, Best Mixed-use Development

· 2014 年 Chicago Athenaeum 国际建筑大奖
——优胜奖
· The Chicago Athenaeum, International Architecture Awards 2014
—Winner

· 2014 年 Cityscape 新兴市场建筑大奖
——已建零售项目优胜奖
· Cityscape Awards for Architecture in Emerging Markets 2014
—Winner, Built Retail Project Award

· 2013—2014 年 A' 设计奖
——建筑，楼宇及结构设计金奖
· A' Design Awards 2013-2014
—Golden Winner, Architecture, Building and Structure Design

· 2013 年香港建筑师学会年奖
——境外商业建筑优异奖
· The Hong Kong Institute of Architects Annual Awards 2013
—Merit Award, Commercial Building Outside Hong Kong

· 2013 年 CIHAF 设计中国大奖
——Mall 类别优胜奖
· CIHAF Design China Award 2013
—Winner, Mall

· 2012 年亚太房地产大奖
——亚太最佳综合建筑优胜奖
——中国最佳综合建筑 5 星奖
· Asia Pacific Property Awards 2012
—Winner, Best Mixed-use Architecture, Asia Pacific
—5-Star, Best Mixed-use Architecture, China

No.14 苏州西交利物浦大学中心楼
Xi'an Jiaotong-Liverpool University Central Building, Suzhou

· 2017 年 AAP 建筑大奖
——建筑设计教育项目荣誉奖
· AAP Architecture Prize 2017
—Honorable Mention, Architectural Design Educational Buildings

· 2016 IDA 国际设计奖
——专业建筑—公共机构建筑项目荣誉奖
· IDA International Design Awards 2016
—Honorable Mention, Professional Architecture, Institutional

· 2015 年亚洲最具影响力设计大奖
——亚洲最具影响力设计铜奖
· Design For Asia Award (DFAA) 2015
—DFA Bronze Award

· 2015 年香港建筑师学会两岸四地建筑设计大奖
——商业办公大楼组别金奖
· HKIA Cross-strait Architectural Design Awards 2015
—Gold Award, Commercial: Office

· 2014 年 MIPIM ASIA 大奖
——特别评审大奖
· MIPIM Asia Awards 2014
—Special Jury Award

· 2014 年亚洲建 + 设大奖
——公共机构 / 公共空间〈专业〉建筑优异奖
· A&D Trophy Awards 2014
—Certificate of Excellence, Architecture (Professional) Institutional / Public Space

· 2014 年南华早报 Chivas 18 建筑设计年奖
——年度大奖
——中国公共 / 社区建筑优胜奖
· SCMP Chivas 18 Architecture and Design Awards 2014
—Grand Prize
—Winner, Public / Community Building, Greater China

· 2013—2014 年 A' 设计奖
——建筑，楼宇及结构设计白金奖
· A' Design Awards 2013-2014
—Platinum Winner, Architecture, Building and Structure Design

· 2012 年 MIPIM ASIA 大奖
——最佳中国未来建设项目金奖
· MIPIM Asia Awards 2012
—Gold Winner, Best Chinese Futura Project

· 2011 全国人居经典规划设计方案竞赛
——最佳设计方案金奖
· China's Outstanding Architectural Design & Planning Award 2011
—Gold Award, Best Design Award

· 2009 年亚太商业房地产大奖
——中国最佳建筑 4 星奖
· Asia Pacific Commercial Property Awards 2009
—4-Star, Architecture Award, China

NO.15 义乌之心
The Heart of Yiwu, Yiwu

· 2017—2018 年 A' 设计奖
——建筑，楼宇及结构设计金奖
· A' Design Awards 2017-2018
—Golden Winner, Architecture, Building and Structure Design

· 2015 年 MIPIM ASIA 大奖
——最佳中国未来大型建设项目铜奖
· MIPIM Asia Awards 2015
—Bronze Winner, Best Chinese Futura Mega Project

· 2015 年亚太房地产大奖
——中国零售建筑优异奖
· Asia Pacific Property Awards 2015
—Highly Commended, Retail Architecture, China

· 2015 年 MIPIM 《建筑评鉴》杂志未来项目大奖
——零售和休闲优异奖
· MIPIM Architectural Review Future Projects Awards 2015
—Commended, Retail and Leisure category

NO.16 广州南丰商业、酒店及展览综合大楼
Nanfung Commercial, Hospitality and Exhibition Complex, Guangzhou

· 2015 年香港建筑师学会两岸四地建筑设计大奖
——酒店组别卓越奖
· HKIA Cross-strait Architectural Design Awards 2015
—Nominated Award, Hotels

· 2014 年 MIPIM ASIA 大奖
——最佳综合发展项目铜奖
——最佳酒店及旅遊发展项目银奖
· MIPIM Asia Awards 2014
—Bronze, Best Mixed-use Development
—Silver, Best Hotel & Tourism Development

· 2014 年亚洲建 + 设大奖
——综合用途〈专业〉建筑优异奖
· A&D Trophy Awards 2014
—Certificate of Excellence, Architecture (Professional) Mixed-Use

· 2014 年 Chicago Athenaeum 国际建筑大奖
——优胜奖
· The Chicago Athenaeum, International Architecture Awards 2014
—Winner

· 2014 年 Cityscape 新兴市场建筑大奖
——已建综合发展项目优胜奖
· Cityscape Awards for Architecture in Emerging Markets 2014
—Winner, Built Mixed-use Project Award

· 2014 年南华早报 Chivas 18 建筑设计年奖
——中国商业及综合建筑优胜奖
· SCMP Chivas 18 Architecture and Design Awards 2014
—Winner, Commercial & Mixed-use Building, Greater China

· 2013 年 Emporis 摩天大楼奖
——世界十大摩天大楼
· Emporis Skyscraper Award 2013
—World's Top 10 Skyscrapers

· 2013—2014 年 A' 设计奖
——建筑，楼宇及结构设计金奖
· A' Design Awards 2013-2014
—Golden Winner, Architecture, Building and Structure Design

· 2013 年 CIHAF 设计中国大奖
——综合类别优胜奖
· CIHAF Design China Award 2013
—Winner, Mixed-use

· 2013 年香港建筑师学会两岸四地建筑设计大奖
——商场 / 步行街银奖
· HKIA Cross-strait Architectural Design Awards 2013
—Silver Award, Commercial: Shopping Center

· 2011 全国人居经典规划设计方案竞赛
——最佳设计方案金奖
· China's Outstanding Architectural Design & Planning Award 2011
—Gold Award, Best Design Award

· 2011 年 Cityscape 新兴市场建筑大奖
——未来休闲项目优胜奖
——未来商业及综合发展项目入围
· Cityscape Awards for Architecture in Emerging Markets 2011
—Winner, Leisure Future Projects
—Finalist, Commercial / Mixed-use Future Projects

· 2010 年亚太商业房地产大奖
——中国综合建筑优异奖
· Asia Pacific Commercial Property Awards 2010
—Highly Commended, Mixed-use Architecture Award, China

NO.17 广东邦华环球贸易中心
Bravo Park Place, Guangzhou

· 2014 年国际房地产大奖
——国际最佳商业高层建筑优胜奖
· International Property Awards 2014
—Winner, Best International Commercial High-rise Architecture

· 2014 年亚太房地产大奖
——亚太最佳商业高层建筑优胜奖
——中国最佳商业高层建筑 5 星奖
· Asia Pacific Property Awards 2014
—Winner, Best Commercial High-rise Architecture, Asia Pacific
—5-Star, Best Commercial High-rise Architecture, China

NO.18 珠海粤澳合作中医药科技产业园总部大楼
Headquarters, Traditional Chinese Medicine Science and Technology Industrial Park of Co-operation between Guangdong and Macao, Zhuhai

· 2017 年香港建筑师学会两岸四地建筑设计大奖
——未兴建项目：建筑方案设计组别卓越奖
· HKIA Cross-strait Architectural Design Awards 2017
—Nominated Award, Un-built Projects: Architectural Design Scheme

· 2016—2017 年 A' 设计奖
——建筑，楼宇及结构设计金奖
· A' Design Awards 2016-2017
—Golden Winner, Architecture, Building and Structure Design

· 2016 年国际房地产大奖
——国际最佳办公楼建筑优胜奖
· International Property Awards 2016
—Winner, Best International Office Architecture

· 2016 年亚太房地产大奖
——中国最佳办公楼建筑 5 星奖
——亚太最佳办公楼建筑优胜奖
· Asia Pacific Property Awards 2016
—5-Star, Best Office Architecture, China
—Winner, Best Office Architecture, Asia Pacific

· 2015 年 Cityscape 新兴市场建筑大奖
——未来商业项目优胜奖
· Cityscape Awards for Architecture in Emerging Markets 2015
—Winner, Future Commercial Project Award

NO.20 珠海横琴国际金融中心
Hengqin International Financial Center, Zhuhai

· 2017 年 AAP 建筑大奖
——建筑设计高层建筑优胜奖
——年度建筑设计奖
· AAP Architecture Prize 2017
—Winner, Architectural Design Tall Buildings
—Architectural Design of the Year

· 2016—2017 年 A' 设计奖
——建筑，楼宇及结构设计金奖
· A' Design Awards 2016-2017
—Golden Winner, Architecture, Building and Structure Design

· 2015 年 MIPIM ASIA 大奖
——最佳中国未来建设项目铜奖
· MIPIM Asia Awards 2015
—Bronze Winner, Best Chinese Futura Project

· 2014 年亚太房地产大奖
——中国综合建筑优异奖
· Asia Pacific Property Awards 2014
—Highly Commended, Mixed-use Architecture, China

NO.21 珠海横琴中冶总部大厦（二期）
Hengqin MCC Headquarters Complex (Phase II), Zhuhai

· 2018 年全球 RLI 大奖
——RLI 未来项目优胜奖
· Global RLI Awards 2018
—Winner, RLI Future Project

· 2017—2018 年 A' 设计奖
——建筑，楼宇及结构设计金奖
· A' Design Awards 2017 - 2018
—Golden Winner, Architecture, Building and Structure Design

· 2017 年 AAP 建筑大奖
——建筑设计高层建筑荣誉奖
· AAP Architecture Prize 2017
—Honorable Mention, Architectural Design Tall Buildings

NO.22 香港西九龙站
Hong Kong West Kowloon Station, Hong Kong

· 2016—2017 年 A' 设计奖
——建筑，楼宇及结构设计金奖
· A' Design Awards 2016-2017
—Golden Winner, Architecture, Building and Structure Design

· 2016 年建筑大奖
——交通项目铜奖
· APP Architecture Prize 2016
—Bronze Winner, Transportation

· 2015 年 MIPIM ASIA 大奖
——最佳未来大型建设项目金奖
· MIPIM Asia Awards 2015
—Gold Winner, Best Futura Mega Project

· 2015 年欧洲领先建筑师论坛大奖
——年度最佳未来建设项目（建设中）优胜奖
· Leading European Architects Forum (LEAF) Awards 2015
—Winner, Best Future Building of the Year-Under Construction

· 2015 年 Chicago Athenaeum 国际建筑大奖
——优胜奖
· The Chicago Athenaeum, International Architecture Awards 2015
—Winner

· 2015 年 MIPIM《建筑评鉴》杂志未来项目大奖
——大城市项目优异奖
· MIPIM Architectural Review Future Projects Awards 2015
—High Commendation, Big Urban Projects

· 2012 年 MIPIM 大奖
——最佳未来大型建设项目
· MIPIM Awards 2012
—Best Futura Mega Project

· 2011 全国人居经典规划设计方案竞赛
——最佳设计方案金奖
· China's Outstanding Architectural Design & Planning Award 2011
—Gold Award, Best Design Award

· 2010 年 Cityscape 新兴市场建筑大奖
——未来旅游，旅行 及交通项目优胜奖
——未来商业及综合发展项目优异奖
· Cityscape Awards for Architecture in Emerging Markets 2010
—Winner, Future Tourism, Travel & Transport Projects
—Highly Commended, Future Commercial / Mixed-use Projects

· 2010 年世界建筑节大奖
——未来基建项目优胜奖
· World Architecture Festival (WAF) Awards 2010
—Winner, Future Infrastructure Projects

NO.24 香港国际机场中场客运廊
Hong Kong International Airport Midfield Concourse, Hong Kong

· 2016 年 ENR 全球最佳项目大奖
——全球最佳机场 / 港口项目优胜奖
· ENR Global Best Projects Awards 2016
—Winner, Global Best Project, Airport / Port

· 2016 年环保建筑大奖
——新建建筑类别已落成建筑 - 公用建筑类大奖
· Green Building Award 2016
—Grand Prize, New Buildings: Completed Projects – Institutional Building

· 2013 年 Autodesk Hong Kong BIM Awards 大奖
——优胜奖
· Autodesk Hong Kong BIM 2013
—Winner

· 2012 年环保建筑大奖
——新建建筑类别（设计中建筑）- 香港优异奖
· Green Building Award 2012
—Merit Award, New Building Category (Building Project under Design) - Hong Kong

· 2011 年 Be Inspired BIM 大奖
——校园、机场及军用设备之创意设计优胜奖
· Be Inspired BIM Awards 2011
—Winner, Innovation in Campuses, Airports & Military Installations

NO.26 香港富临阁
The Forum, Hong Kong

· 2017 年 AAP 建筑大奖
——建筑设计商业建筑荣誉奖
· AAP Architecture Prize 2017
—Honorable Mention, Architectural Design Commercial Architecture

· 2016 年 ARCASIA 建筑大奖
——公众设施商业大厦优胜奖
· ARCASIA Awards for Architecture 2016
—Mention, Public Amenity: Commercial Buildings

· 2015 年亚洲建 + 设大奖
——商业，零售及办公楼〈专业〉建筑优异奖
· A&D Trophy Awards 2015
—Certificate of Excellence, Architecture (Professional) Commercial, Retail or Office

· 2015 年 Cityscape 新兴市场建筑大奖
——已建商业项目优胜奖
· Cityscape Awards for Architecture in Emerging Markets 2015
—Winner, Built Commercial Project Award

· 2015 年 Chicago Athenaeum 国际建筑大奖
——优胜奖
· The Chicago Athenaeum, International Architecture Awards 2015
—Winner

· 2015 年南华早报 Chivas 18 建筑设计年奖
——香港商业及综合体年度建筑优胜奖
· SCMP Chivas 18 Architecture and Design Awards 2015
—Winner, Commercial & Mixed-use Architect of the Year (Hong Kong)

· 2015 年香港建筑师学会两岸四地建筑设计大奖
——商业办公大楼组别卓越奖
· HKIA Cross-strait Architectural Design Awards 2015
—Nominated Award, Commercial: Office

· 2014 年香港建筑师学会年奖
——境内商业建筑优异奖
· The Hong Kong Institute of Architects Annual Awards 2014
—Merit Award, Commercial Building of Hong Kong

· 2014 年亚太房地产大奖
——亚太最佳办公楼建筑优胜奖
——香港最佳办公楼建筑 5 星奖
· Asia Pacific Property Awards 2014
—Winner, Best Office Architecture, Asia Pacific
—5-Star, Best Office Architecture, Hong Kong

NO.27 台北砳建筑
Lè Architecture, Taipei

· 2018 年 Architizer A+ 大奖
——高层（16 层 +）商业办公大楼最受欢迎奖
· Architizer A+ Awards 2018
—Popular Choice Winner, Commercial-Office - High Rise (16+ Floors)

· 2017—2018 年 A' 设计奖
——建筑，楼宇及结构设计金奖
· A' Design Awards 2017-2018
—Golden Winner, Architecture, Building and Structure Design

· 2017 年国际房地产大奖
——国际最佳办公楼建筑优胜奖
· International Property Awards 2017
—Winner, Best International Office Architecture

· 2017 年亚太房地产大奖
——亚太办公楼建筑优胜奖
——台湾办公楼建筑 5 星奖
· Asia Pacific Property Awards 2017
—Winner, Best Office Architecture, Asia Pacific
—5-Star, Office Architecture, Taiwan

NO.28 台中商业银行企业总部综合项目
Commercial Bank Headquarters Mixed-use Project, Taichung

· 2017—2018 年 A' 设计奖
——建筑，楼宇及结构设计金奖
· A' Design Awards 2017-2018
—Golden Winner, Architecture, Building and Structure Design

· 2018 年 MIPIM《建筑评鉴》杂志未来项目大奖
——高层建筑优胜奖
· MIPIM Architectural Review Future Project Awards 2018
—Winner, Tall Buildings category

NO.29 新加坡星宇项目
The Star, Singapore

· 2017 年 AAP 建筑大奖
——建筑设计综合建筑荣誉奖
——建筑设计文化项目荣誉奖
· AAP Architecture Prize 2017
—Honorable Mention, Architectural Design Mixed-use Architecture
—Honorable Mention, Architectural Design Cultural Architecture

· 2016 年 IDA 国际设计奖
——专业建筑—新商业建筑项目荣誉奖
· IDA International Design Awards 2016
—Honorable Mention, Professional Architecture, New Commercial Building

· 2014 年亚洲建 + 设大奖
——综合用途〈专业〉建筑优异奖
——商业，零售及办公楼〈专业〉建筑优异奖
· A&D Trophy Awards 2014
—Certificate of Excellence, Architecture (Professional) Mixed-Use
—Certificate of Excellence, Architecture (Professional) Commercial, Retail or Office

· 2014 年 ENR 全球最佳项目大奖
——全球最佳零售 / 综合发展项目优胜奖
· ENR Global Best Projects Awards 2014
—Winner, Best Global Project, Retail / Mixed-use Developments

· 2014 年优质建筑大奖
——境外建筑项目优异奖
· Quality Building Award 2014
—Merit Award, Building Outside Hong Kong

· 2013—2014 年 A' 设计奖
——建筑，楼宇及结构设计金奖
· A' Design Awards 2013-2014
—Golden Winner, Architecture, Building and Structure Design

· 2013 年 MIPIM ASIA 大奖
——最佳零售休闲发展项目银奖
· MIPIM Asia Awards 2013
—Silver, Best Retail and Leisure Development

· 2010 年 MIPIM《建筑评鉴》杂志未来项目大奖
——综合发展项目优异奖
· MIPIM Architecture Review Future Project Awards 2010
—Commended, Mixed-use Projects

· 2010 年 Cityscape 新兴市场建筑大奖
——未来商业及综合发展项目优胜奖
· Cityscape Awards for Architecture in Emerging Markets 2010
—Winner, Future Commercial / Mixed-use Projects

· 2009 年亚太商业房地产大奖
——亚太最佳建筑优胜奖
——新加坡最佳建筑 5 星奖
· Asia Pacific Commercial Property Awards 2009
—Winner, Architecture Award, Asia Pacific
—5-Star, Architecture Award, Singapore

NO.30 新加坡 Sandcrawler
Sandcrawler, Singapore

· 2017 年美国建筑师学会国际地区设计大奖
——公开国际建筑设计荣誉奖
· AIA International Region Design Awards 2017
—Honor Award, Open International Architecture

· 2016—2017 年 A' 设计奖
——建筑，楼宇及结构设计银奖
· A' Design Awards 2016-2017
—Silver Winner, Architecture, Building and Structure Design

· 2016 年 AAP 建筑大奖
——建筑设计商业建筑荣誉奖
——景观建筑商业项目荣誉奖
· AAP Architecture Prize 2016
—Honorable Mention, Architectural Design Commercial Architecture
—Honorable Mention, Landscape Architecture Commercial

· 2015 年 ARCASIA 建筑大奖
——公众设施商业大厦优胜奖
· ARCASIA Awards for Architecture 2015
—Mention, Public Amenity: Commercial Buildingse

· 2014 年香港建筑师学会年奖
——境外商业建筑优异奖
· The Hong Kong Institute of Architects Annual Awards 2014
—Merit Award, Commercial Building Outside Hong Kong

· 2014 年 MIPIM ASIA 大奖
——最佳办公楼及商业发展项目金奖
· MIPIM Asia Awards 2014
—Gold, Best Office & Business Development

· 2014 年亚洲建 + 设大奖
——商业，零售及办公楼〈专业〉建筑优异奖
· A&D Trophy Awards 2014
—Certificate of Excellence, Architecture (Professional) Commercial, Retail or Office

· 2014 年新加坡总统设计奖
——年度设计优胜奖
· President's Design Award Singapore 2014
—Winner, Design of the Year

· 2014 年 AIA 西北及太平洋地区设计大奖
——荣誉奖
· 2014 AIA Northwest and Pacific Region Design Awards
—Honor Award

· 2011 年 MIPIM ASIA 大奖
——最佳未来建设项目金奖
· MIPIM Asia Awards 2011
—Gold Winner, Best Futura Projects

· 2011 年 Cityscape 新兴市场建筑大奖
——未来商业及综合发展项目优胜奖
· Cityscape Awards for Architecture in Emerging Markets 2011
—Winner, Future Commercial/ Mixed-use Projects

· 2011 年亚太商业房地产大奖
——新加坡最佳办公楼建筑 5 星奖
· Asia Pacific Commercial Property Awards 2011
—5-Star, Best Office Architecture Award, Singapore

· 2011 年 Chicago Athenaeum 国际建筑大奖
——最佳全球新设计
· The Chicago Athenaeum, International Architecture Awards 2011
—Best New Global Design

NO.31 阿联酋迪拜 Ocean Heights
Ocean Heights, Dubai, UAE

· 2005 年 Bentley Design 国际房地产大奖
——最佳建筑大奖
· Bentley Design International Property Award 2005
—Best Architecture

附录 2 APPENDIX 2

SELECTED PROJECTS

精选项目

综合体 MIXED-USE

Blekerij，比利时科特赖克
Blekerij, Kortrijk, Belgium

新华书店集团公司解放碑时尚文化城，中国重庆
Xinhua Bookstore Group Jiefangbei Book City Mixed-use Project, Chongqing, PRC

P 084

中洲湾 C FutureCity，中国深圳
C FutureCity, Shenzhen, PRC

无锡恒隆广场，中国无锡
Center 66, Wuxi, PRC

P 112

吉盟国际大厦，中国深圳
Gmond International Building, Shenzhen, PRC

恒大广场，中国成都
Evergrande Plaza, Chengdu, PRC

P 078

武汉恒隆广场，中国武汉
Heartland 66, Wuhan, PRC

P 090

台中商业银行企业总部项目，中国台中
Commercial Bank Headquarters Mixed-use Project, Taichung, PRC

P 202

横琴国际金融中心，中国珠海
Hengqin International Financial Center, Zhuhai, PRC

P 150

北京大兴 3 及 4 地块项目，中国北京
Daxing Plots 3 and 4, Beijing, PRC

P 056

财富广场第一期，中国北京
Fortune Plaza Phase 1, Beijing, PRC

金地商置长寿路项目，中国上海
Gemdale Changshou Road, Shanghai, PRC

仁和春天国际广场及国际花园，中国成都
Renhe Spring Mixed-use Development, Chengdu, PRC

绿地东村项目 CBED 地块，中国成都
Greenland East Village CBED Plots, Chengdu, PRC

深圳罗湖友谊城，中国深圳
Shenzhen Luohu Friendship Trading Center, Shenzhen, PRC

绿地双鹤湖双塔项目，中国郑州
Greenland Zhengzhou Double-Crane Lake Mixed-use Development, Zhengzhou, PRC

SFC 协信中心，中国重庆
Sincere Financial Center, Chongqing, PRC

珠海横琴中冶总部大厦（二期），中国珠海
Hengqin MCC Headquarters Complex (Phase II), Zhuhai, PRC

P 158

虹桥世界中心，中国上海
Hongqiao World Center, Shanghai, PRC

P 108

U-Bora 综合开发项目，阿联酋迪拜
U-Bora Towers, Dubai, UAE

龙湖虹桥项目，中国上海
Longfor Hongqiao Mixed-use Project, Shanghai, PRC

P 102

南丰商业、酒店及展览综合大楼，中国广州
Nanfung Commercial, Hospitality and Exhibition Complex, Guangzhou, PRC

P 130

徐州苏宁广场，中国徐州
Xuzhou Suning Plaza, Xuzhou, PRC

徐州华厦广场，中国徐州
Xuzhou Huasha Plaza, Xuzhou, PRC

北苑北辰综合开发项目，中国北京
North Star Mixed-use Development, Beijing, PRC

P 068

珠海横琴新区 54# 地块项目，中国珠海
Zhuhai Hengqin CRCC Plaza Project, Zhuhai, PRC

Abdul Latif Jameel 总部大厦，沙特阿拉伯吉达
Abdul Latif Jameel's Corporate Headquarters, Jeddah, Saudi Arabia

Boulevard Plaza，阿联酋迪拜
Boulevard Plaza, Dubai, UAE

新浪总部大楼，中国北京
Sina Plaza, Beijing, PRC

P 060

重庆高科北部之光办公楼项目，中国重庆
Chongqing Gaoke Group Ltd Office Project, Chongqing, PRC

北京大望京综合开发项目，中国北京
Da Wang Jing Plot#2 Mixed-use Development, Beijing, PRC

P 050

珠海粤澳合作中医药科技产业园总部大楼，中国珠海
Headquarters, Traditional Chinese Medicine Science and Technology Industrial Park of Co-operation between Guangdong and Macao, Zhuhai, PRC

P 142

嘉华上海闸北综合体项目，中国上海
K.Wah Zhabei Mixed-use Shanghai, Shanghai, PRC

砳建筑，中国台北
Lè Architecture, Taipei, PRC

P 196

邦华环球贸易中心，中国广州
Bravo Park Place, Guangzhou, PRC

P 136

方正国际金融中心，中国武汉
Founder International Financial Center, Wuhan, PRC

北苑北辰综合开发项目，中国北京
North Star Mixed-use Development, Beijing, PRC

P 068

南沙建滔自贸区综合体项目，中国广州
Nansha Kingboard Free Trade Zone Mixed-use Project, Guangzhou, PRC

NTC 贸易中心，中国台中
National Trade Center, Taichung, PRC

富力中心，中国广州
R&F Center, Guangzhou, PRC

Sandcrawler，新加坡
Sandcrawler, Singapore

P 216

上海世博会地区 A13A-01 地块新建营业办公楼项目，中国上海
Shanghai EXPO Plot A13A-01 Project, Shanghai, PRC

上海星荟中心，中国上海
Shanghai Landmark Center, Shanghai, PRC

P 096

建滔广场二期，中国上海
The Bankside Towers, Shanghai, PRC

上海张江创意企业综合服务平台项目，中国上海
Shanghai Zhangjiang R&D Park Office, Shanghai, PRC

首建嘉定金融中心，中国上海
Shoujian Jiading Financial Center, Shanghai, PRC

上海 K2 新华路项目，中国上海
Shanghai K2 Xinhua Lu Project, Shanghai, PRC

上海新富港金融中心 1 座，中国上海
Shanghai New Rich Port Center Tower 1, Shanghai, PRC

联合利华总部大楼，印度尼西亚雅加达
Unilever Headquarters, Jakarta, Indonesia

富临阁，中国香港
The Forum, Hong Kong, PRC

P 192

郑州正弘航空港区办公楼项目，中国郑州
Zhenghong Property Air Harbour Office Project, Zhengzhou, PRC

商业零售 RETAIL

香港中环区购物中心，中国香港
Centers of Central, Hong Kong, PRC

星光时代广场，中国重庆
Starlight Place, Chongqing, PRC

义乌之心，中国义乌
The Heart of Yiwu, Yiwu, PRC

P 126

MOKO 新世纪广场，中国香港
MOKO, Hong Kong, PRC

青岛金茂湾购物中心，中国青岛
Jinmao Harbour Shopping Center, Qingdao, PRC

P 072

星宇项目，新加坡
The Star (Star Vista), Singapore

P 210

大连恒隆广场，中国大连
Olympia 66, Dalian, PRC

P 042

研究和制药设施 HEALTHCARE & RESEARCH

Celgene Seattle，美国西雅图
Celgene Seattle, Seattle, USA

传染病研究所，美国西雅图
Infectious Disease Research Institute, Seattle, USA

西雅图儿童研究所 Building Cure，美国西雅图
Seattle Children's Research Institute - Building Cure, Seattle, USA

西雅图儿童研究所西部 8 号，美国西雅图
Seattle Children's Research Institute - West 8th, Seattle, USA

国家新能源汽车动力电池及驱动系统质量监督检测中心，中国珠海
National New Energy Vehicle Power Battery and Electric Drive System Quality Supervision and Inspection Center, Zhuhai, PRC

酒店 HOTEL

Alacarte 公寓酒店，越南下龙湾
Alacarte Condotel, Ha Long City, Vietnam

港岛英迪格酒店，中国香港
Hotel Indigo Hong Kong Island, Hong Kong, PRC

吉林金融中心五星级酒店项目，中国吉林
Jilin Financial Center 5-Star Hotel Project, Jilin, PRC

石狮黄金海岸项目，中国石狮
Shishi Gold Coast Project, Shishi, PRC

皇冠假日酒店，中国惠州
Crowne Plaza Hotel, Huizhou, PRC

上海建滔诺富特酒店，中国上海
Novotel Shanghai Hongqiao Hotel, Shanghai, PRC

碧荟，中国香港
The Beacon, Hong Kong, PRC

成都瑞吉酒店，中国成都
The St. Regis Chengdu, Chengdu, PRC

伦敦圣潘克拉斯万丽酒店，英国伦敦
St. Pancras Renaissance Hotel, London, UK

台北 Panco 酒店，中国台北
Taipei Panco, Taipei, PRC

Vida Dubai Mall，阿联酋迪拜
Vida Dubai Mall, Dubai, UAE

珠海横琴天湖酒店项目，中国珠海
Zhuhai Hengqin Tianhu Hotel Development, Zhuhai, PRC

住宅 RESIDENTIAL

DAMAC Heights，阿联酋迪拜
DAMAC Heights, Dubai, UAE

Ocean Heights，阿联酋迪拜
Ocean Heights, Dubai, UAE

P 222

臻环，中国香港
Gramercy, Hong Kong, PRC

阿拉姆舒特拉公寓，印度尼西亚雅加达
Alam Sutera Residence, Jakarta, Indonesia

Asimont 别墅，新加坡
Asimont Villas, Singapore

仁和春天国际花园，中国成都
Renhe Spring Residential Development, Chengdu, PRC

鹏利海景一号，中国上海
TG Harbour View Apartment, Shanghai, PRC

Regent Quarter C 区及 D 区，英国伦敦
Regent Quarter Block C & D, London, UK

浅水湾地段第 1165 号住宅项目，中国香港
R.B.L 1165 Repulse Bay, Hong Kong, PRC

THR350 私人寓所，中国香港
THR350, Hong Kong, PRC

苍鹭大厦，英国伦敦
The Heron, London, UK

基础设施 INFRASTRUCTURE

迪拜地铁站，阿联酋迪拜
Dubai Metro, Dubai, UAE

P 226

香港国际机场北卫星客运廊，中国香港
Hong Kong International Airport North Satellite Concourse, Hong Kong, PRC

P 186

厦门港国际旅游客运码头，中国厦门
Xiamen International Cruise Terminal, Xiamen, PRC

滨海湾地铁站，新加坡
Marina Bay MRT Station, Singapore

香港西九龙站，中国香港
Hong Kong West Kowloon Station, Hong Kong, PRC

P 164

港珠澳大桥香港口岸旅检大楼，中国香港
Hong Kong-Zhuhai-Macao Bridge Hong Kong Port - Passenger Clearance Building, Hong Kong, PRC

P 172

香港国际机场中场客运廊，中国香港
Hong Kong International Airport Midfield Concourse, Hong Kong, PRC

P 180

香港铁路迪士尼支线欣澳站，中国香港
MTR Disneyland Resort Line Sunny Bay Station, Hong Kong, PRC

纬壹地铁站，新加坡
One-north MRT Station, Singapore

烟台国际机场 T2 航站楼，中国烟台
Yantai International Airport Terminal 2, Yantai, PRC

深圳宝安国际机场卫星厅，中国深圳
Shenzhen Airport Satellite Concourse, Shenzhen, PRC

P 146

教育设施 EDUCATION

台北欧洲学校阳明山校区校园扩建工程，中国台北
Taipei European School Yangmingshan Campus Redevelopment Project, Taipei, PRC

西交利物浦大学中心楼，中国苏州
Xi'an Jiaotong-Liverpool University Central Building, Suzhou, PRC

P 118

圣保罗男女中学重建，中国香港
St. Paul's Co-educational College, Hong Kong, PRC

香港中文大学图书馆扩建项目，中国香港
The Chinese University of Hong Kong Library Extension, Hong Kong, PRC

Winspear Completion Project, 加拿大埃德蒙顿
Winspear Completion Project, Edmonton, Canada

庞万伦学生中心，中国香港
Pommerenke Student Center, Hong Kong, PRC

重建项目 RESTORATION

伦敦大剧院，英国伦敦
London Coliseum, London, UK

黄大仙中心，中国香港
Wong Tai Sin Temple Mall, Hong Kong, PRC

特鲁里街皇家歌剧院，英国伦敦
Theatre Royal, Drury Lane, London, UK

威尔士王子剧院，英国伦敦
Prince of Wales Theatre, London, UK

Royal & Derngate，英国北安普顿
Royal & Derngate, Northampton, UK

艺术社区：湾仔茂萝街 / 巴路士街活化及文物保育项目，中国香港
Art Community Revitalization Project at Mallory Street Burrow Street, Wan Chai, Hong Kong, PRC

会议及展览设施 CONVENTION & EXHIBITION

华南城展览中心，中国南宁
Huanancheng Exhibition Center, Nanning, PRC

小型项目 SMALL PROJECTS

昆泰望京二号展示馆，中国北京
Kuntai Wangjing Plot 2 Exhibition Center, Beijing, PRC

中欧国际城展览中心，中国青岛
Qingdao Sino EU International City Exhibition Center, Qingdao, PRC

莱安体验中心，中国西安
LAND Experience Center, Xi'an, PRC

城市设计及总体规划 URBAN DESIGN AND MASTERPLANNING

广州国际金融城，中国广州
Guangzhou International Financial City, Guangzhou, PRC

Project (Re) Plant，马来西亚婆罗洲
Project (Re) Plant, Borneo, Malaysia

嘉定新城总部园区，中国上海
Jiading New Town Headquarters Park, Shanghai, PRC

南京江北金融中心一期，中国南京
Nanjing Jiangbei Financial Center Phase I, Nanjing, PRC

武汉昆仑城总体规划，中国武汉
Wuhan Kunlun City Master Plan, Wuhan, PRC

艺术及休闲设施 ARTS & LEISURE

萨德勒斯威尔士剧院，英国伦敦
Sadler's Wells Theatre, London, UK

布里奇沃特音乐大厅，英国曼彻斯特
The Bridgewater Hall, Manchester, UK

艺术及休闲设施 ARTS & LEISURE

艾尔斯伯里水滨剧院，英国艾尔斯伯里
Aylesbury Waterside Theatre, Aylesbury, UK

市政厅音乐及戏剧学院，英国伦敦
Guildhall School of Music and Drama, London, UK

中国国际贸易中心三期 C 阶段发展项目，中国北京
China World Trade Center Phase 3C Development, Beijing, PRC

Cast 剧院，英国唐卡斯特
Cast, Doncaster, UK

P 232

北方剧场，英国纽卡斯尔
Northern Stage, Newcastle, UK

Tara 艺术剧院，英国伦敦
Tara Theatre, London, UK

超高层 SUPER HIGH RISE

DAMAC Heights，阿联酋迪拜
DAMAC Heights, Dubai, UAE

珠海横琴国际金融中心，中国珠海
Hengqin International Financial Center, Zhuhai, PRC

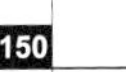

P 150

珠海横琴中冶总部大厦（二期），中国珠海
Hengqin MCC Headquarters Complex (Phase II), Zhuhai, PRC

P 158

Ocean Heights，阿联酋迪拜
Ocean Heights, Dubai, UAE

P 222

长沙金茂超高层项目，中国长沙
Changsha Jinmao Super High Rise Project, Changsha, PRC

天山·世界之门 27 及 28 号地，中国石家庄
Tianshan Gate of the World Plots 27 and 28, Shijiazhuang, PRC

武汉恒隆广场，中国武汉
Heartland 66, Wuhan, PRC

P 090

后 记
Afterword

文 支文军 by ZHI Wenjun

《Aedas 在中国》是本聚焦 Aedas 成长与发展的图书。Aedas 作为国际化的全球十大设计公司之一，在中国城市化发展实践中积累了丰富的经验，也给城市留下了瞩目的作品。中国城市的快速发展与进步，激励着 Aedas 完善其世界多元化的理念，确立其伟大的设计是环境的产物的设计观。这种多样性和创作性为城市带来了创新的项目和具有国际水准的高品质建筑环境。这些理念与作品在此 30 多年间沉淀孕育，逐步形成了 Aedas 的独特思想，有对全球化、城市化的前瞻，有对网络化、新科技的运用，有对中国城市建筑与运行高效的设想，同时也有对文化的关注和对走向全球化时代“一带一路”倡议的回应。Aedas 积极投身公益教育，分别携手内地知名建筑高校如同济大学及清华大学，创办建筑奖学金及创新基金，鼓励建筑学院的学子们坚持对建筑的热忱，力所能及地去帮助到这些未来的建筑师们在追求建筑理想的道路上减少一些压力。这让我们看到 Aedas 的社会担当。

本书的编著过程是一份特殊的经历。我们南下香港、北上北京，对 Aedas 主席和董事们进行了访谈，在交流的过程中，我们总能感受到 Aedas 对一个多元化缤纷多彩世界的理想，他们身上有对创作的渴望、有独立的思考与执着的精神。我们了解到 Aedas 创造的是一个全球性优秀设计创意的平台，集结了世界各地极具天赋的建筑师。我们还实地走访考察了 Aedas 的建筑项目，通过与专业摄影的互动，更好地去呈现建成项目的价值。Aedas 的项目，从城市文化、建成环境到空间结构、材料细节等，都让我们能从中感受到设计对地域、环境、文化差异的理解，对使用群体的充分尊重，无论是中国香港、深圳、珠海，还是上海、苏州、重庆……这些卓越的设计解决方案，为无数城市增添了光彩。

特别感谢郑时龄院士和 Aedas 主席纪达夫先生为本书写序。感谢在图书编著过程中得到 Aedas 团队耐心细致的辅助和大量沟通协调的积极配合。特别感谢冯仕达、吕炜对本书出版的支持。感谢苏杭、郭小溪、费甲辰等在本书制作过程中的辛勤劳动。感谢完颖对本书的装帧设计和排版。感谢摄影师张虔希提供的专业图片。感谢《时代建筑》编辑部以及同济大学出版社的同仁给予的支持。再次对参与本书的所有工作人员表示由衷的感谢。在既繁琐又辛苦的编辑过程中，大家齐心协力，历经反复修改和确认，研究与编撰成果才得以呈现。

Aedas in China is a book focusing on the growth and development of Aedas. As one of the top ten global design companies, Aedas has accumulated rich experience in dealing with China's urbanisation, leaving behind its marks in the country's fast-growing cities. The rapid development and progress of Chinese cities inspire Aedas to improve its vision to be diverse and to confirm that great designs are products of their environment. Such diversity and creativity give cities innovative projects and a high-quality built environment that meets the highest international standards. Over the past two decades, this vision and these works have gradually formed Aedas' unique ideology, which embraces globalisation and urbanisation, along with application of the internet and other new technologies, a concern for Chinese cities and their architecture, a concern for culture as well as a response to the globally-focused Belt and Road Initiative. Aedas is also actively engaged in education, working hand in hand with renowned architecture schools including Tongji University and Tsinghua University to establish scholarships and innovation fund to encourage architecture students to keep their passion with less pressure and more help. Aedas will certainly shoulder even more social responsibility in the future.

The editing process of this book has been a special experience. Traveling south to Hong Kong and then north to Beijing, we interviewed Aedas' Chairman and Directors. From our talk we could always sense Aedas' ideal of a diverse and colourful world. They have desire to create an independent way of thinking and an ethos of perseverance. It has occurred to us that Aedas has created a global platform for design excellence and brought together highly talented architects from all over the world. We have also visited Aedas' projects and interacted with professional photographers so as to better present the value of the built projects. From city culture, site context to spatial structures and materials, we experienced designs that relate well to their location, context and cultural environments, all while respecting their users. No matter if they are in Hong Kong, Shenzhen, Zhuhai, Shanghai, Suzhou, Chongqing or any other Chinese cities, these excellent design strategies and proposals add glory to numerous urban areas across the country.

Special thanks to ZHENG Shiling (member of the Chinese Academy of Sciences) and Keith GRIFFITHS (chairman of Aedas) who have written prologues for this book. We appreciate the support from the Aedas team. We would also especially thank Stanislaus FUNG and LYU Wei for their support for the publication of this book. Thanks to SU Hang, GUO Xiaoxi, FEI Jiachen who have also contributed a lot in the process of bookmaking; WAN Ying for the design and typesetting of the book; and ZHANG Qianxi for taking professional pictures. Meanwhile, we appreciate the support from the editorial department of *Times Architecture* and our colleagues from Tongji University Press. Again, we would like to express heartfelt thanks to everyone who has contributed to this book. In the tireless and painstaking editing process, we worked together to present results of research and compilation after repeated modification and confirmation.

图书在版编目（CIP）数据

Aedas 在中国 / 支文军，徐洁著 . -- 上海 : 同济大学出版社，2018.12
ISBN 978-7-5608-7470-8

Ⅰ . ① A… Ⅱ . ①支… ②徐… Ⅲ . ①建筑设计－作品集－中国－现代 Ⅳ . ① TU206